ANALYSIS OF COMPLEX DISEASES

A Mathematical Perspective

ANALYSIS OF COMPLEX DISEASES

A Mathematical Perspective

Guanyu Wang

CRC Press is an imprint of the
Taylor & Francis Group, an **informa** business

CRC Press
Taylor & Francis Group
6000 Broken Sound Parkway NW, Suite 300
Boca Raton, FL 33487-2742

First issued in paperback 2019

ISBN-13: 978-1-4665-7221-8 (hbk)
ISBN-13: 978-0-367-37900-1 (pbk)

Library of Congress Cataloging-in-Publication Data

Wang, Guanyu (Physicist), author.
Analysis of complex diseases : a mathematical perspective / Guanyu Wang.
p. ; cm.
Includes bibliographical references and index.
ISBN 978-1-4665-7221-8 (hardcover : alk. paper)
I. Title.
[DNLM: 1. Metabolic Diseases--complications. 2. Models, Biological. 3. Phosphatidylinositol 3-Kinases--metabolism. 4. Systems Biology--methods. WD 200]

RB147
616.3'9--dc23 2013038208

Visit the Taylor & Francis Web site at
http://www.taylorandfrancis.com

and the CRC Press Web site at
http://www.crcpress.com

Dedication

To my wife Yanbing Xu

To my parents Jinzhi Pan and Zhuyan Wang

Contents

PART II Molecular Level: What Can Be Provided?

PART III Mathematical Analysis of Complex Diseases

Preface

The world has witnessed an escalation in the incidence of many complex diseases. In the United States, about two-thirds of the population is overweight or obese, with about 25% in the obese category; about one-third of children born after the year 2000 will develop diabetes at some point during their lifetime; about one-third of women and half of men will develop cancer. In China, there are more than 92 million diabetic patients and about 140 million prediabetic persons. The global incidence of type 2 diabetes is estimated to double to 350 million cases by the year 2030, with expenditure related to diabetes estimated to reach $132 billion in the United States alone.

It has been suspected that industrialization and modern lifestyles are important factors contributing to the epidemics of complex diseases. Unfortunately, while science and technology have facilitated agriculture and industrialization very effectively, their contributions to the prevention and treatment of complex diseases have not been impressive. For the past 40 years, the overall rate of cancer mortality has been basically flat, even though there was a quantum leap in our understanding of the glorious details within the cell: from molecular dynamics to interaction networks. Apparently, something is missing to synergize what we know, something that is desperately needed to solve the health care crises of our time.

Many people think that our clinical failure is largely due to the immense complexity inherent in complex diseases. A complex disease is almost always multifactorial, involving many etiology and risk factors operating at multiple (molecular, cellular, organismal, environmental, etc.) levels. On the other hand, disparate diseases may share the same etiology but bifurcate at some point during development. Such immense complexity necessitates detailed biological studies at all these levels, especially the molecular level, at which the essential aberrations occur for almost all the complex diseases. With more and more data being collected, one may hope that the complexity can be gradually unraveled and these diseases can finally be cured. While detailed biological studies are certainly necessary, fundamental thinking and top-down approaches are equally important. Theories and mathematical models can enhance our understanding of biological data, without which one may easily be overwhelmed by the blizzard of molecular details but miss the big picture.

Therefore, we must think fundamentally and seek overarching mechanisms that govern the complex biological phenomena. We all know that big breakthroughs in physics are always fundamental. Abstract concepts such as mass and energy, and the related physical laws, are as beautiful and stimulating today as when they were first developed several centuries ago. Behind these brilliant achievements was intense and deep thinking about fundamental aspects of nature, without which only slow, incremental development can be achieved in individual subfields of physics. For the same reason, abstraction and fundamental thinking may allow us to grasp sophisticated, growing biological knowledge holistically. Such a holistic perspective may tell us which molecular details are more important and which are less.

This is exactly the philosophy this book tries to endorse for the field of complex diseases. Instead of studying one specific disease, we try to find mechanistic commonalities underlying the pathogenesis of several diseases. Given that a single disease is already complex, how can one study multiple diseases all together? Before the onset of symptoms, there was a long period of clinical latency, during which some aberration must have occurred. The aberration may be small, but is essential. In contrast, the late phase of the disease is usually dominated by consequences of the early essentials, manifesting as symptoms (nonessential body reactions). Therefore, one does not have to study the whole spectrum of complexities related to a disease. It is the aim of this book to determine the latent aberration that dictates the entire disease progression.

It is well known that gene mutations or genetic defects can kick off the progression of a disease. However, except for a few special cases, a complex disease involves hundreds or even thousands of gene mutations. It is thus very difficult to determine causal relationship, namely, which genes are essential, which genetic changes are only effects, and which genetic changes are simply random. Even if the essential genes can be identified, it would still be difficult to determine their intricate synergy that leads to a rather robust disease phenotype. Despite tremendous technological developmemts that allow our observations to penetrate to the molecular level, we are still nowhere close to discovering the nature of the aberration, for most complex diseases.

It is here that physical thinking shows its importance. The aberration associated with a disease can be quantified as the deviation from normal. Therefore, a disease should not be considered in isolation, but should be considered together with the healthy state, which is the reference point against which the disease progression can be observed. That is, a disease and the normal state are inherently connected. Because the aberration is *small* during the early latency phase, mathematical analysis (*infinitesimal* approximation, singularity analysis, etc.) can be used to study the disease without losing much precision.

With a little more thinking, we realize that we should focus on normality instead of the specific disease that might be our first interest. Since normality can be disturbed in different ways, one might as well identify all the patterns of deviation around the normality. Each pattern represents a family of disturbances that are qualitatively the same and may correspond to a specific disease. In this way, one can include multiple diseases into the same global picture, in which the diseases are qualitatively different but interconnected entities that bifurcate from normality. Such a holistic perspective could facilitate a deeper understanding of complex diseases and contribute to their prevention and treatment.

The above philosophy encourages global thinking, but does not force one to include all the complexities, nor all diseases, into the same framework. It would be very difficult, if not impossible, to do so. In this book, we will consider only diseases related to cellular energy metabolism, which include cancer, diabetes, and other metabolic syndromes. These diseases are not only devastating, but also have much increased incidence in recent years, as demonstrated by the epidemiological data mentioned in the beginning of this book. We will see that the holistic perspective will not only illuminate the molecular mechanisms of these diseases, but also the evolutionary underpinning of their simultaneous epidemics.

In this book, the global view is not only in terms of considering multiple diseases, but also in terms of the consideration of complexities spanning multiple levels, from individual molecules to their interaction network and then to the organismal level. One may first consider events within a cell, namely, the network of interactions among many kinds of biomolecules. By mathematical modeling, the range of dynamical behaviors of the network can be revealed and related to cellular physiology or pathology. That is, some dynamical behaviors correspond to the normal functional response of the cell and other dynamical behaviors correspond to a variety of cellular aberrations. Because the cell is the fundamental unit of life, studies at the subcellular and cellular levels have already provided great insights into the healthy state and the pathogenesis of various diseases. Nevertheless, the story should not end here. It is interesting to know how responses of individual cells shape the whole-body physiology and, conversely, how a number of competing requirements at the organismal level are reconciled to constrain the way a cell functions. In brief, we aim to integrate information from multiple levels so that a unified theory can be developed, which in turn sheds light on research at the respective levels. In physics, a well-known example is the connection between statistical mechanics and thermodynamics. The former was established from the microscopic perspective and the latter was established from the macroscopic perspective. The holistic consideration of both levels has yielded even more insights. In this book, the holistic consideration leads to the conclusion that some disease phenotypes actually correspond to well-evolved control mechanisms that worked optimally for our ancestors in an environment quite different from ours. Acute changes of environment and lifestyle, caused by industrialization and agriculture, may make the evolved control mechanisms less optimal or even detrimental, contributing to the epidemics of these diseases.

In accordance with the multilevel consideration, this book is divided into three parts. Part I describes normal physiology. For complex organisms such as humans, many competing or even opposing requirements have to be met simultaneously. The body as a whole has to work optimally. This happens primarily at the organismal level. Part II describes the organizing principles of molecular circuits, i.e., how different kinds of biomolecules in the cell should be "wired" in order to provide the correct functional output so that various demands at the organismal level can be met and their possible contradictions can be reconciled. This happens primarily at the subcellular, molecular level. Part III addresses the theme of this book — complex diseases, from a unique perspective that combines information from both the organismal level and the molecular level. Although only a small number of metabolic diseases are addressed, I hope the paradigm used in this book can be useful for the study of other categories of diseases. For example, Alzheimer's disease, Parkinson's disease, and some other neurological diseases may be addressed collectively in terms of this philosophy.

The ideas in the book were developed in a time of great grief, during which several relatives, colleagues, and friends died from the very diseases mentioned in this book. Among them were my mother Jinzhi Pan, my mother-in-law Xunzhi Cai, my uncle Xiaoyun Wang, and Profs. Cornelius Bennholdt and Barry Berman, who were Chairs of the Physics Department at George Washington University. I dedicate this book to them for their heroic struggle against these diseases.

I want to pay special tribute to my wife, Yanbing Xu, who has been so tolerant of my rather excessive research efforts and supported them wholeheartedly; to my parents Zhuyan Wang and Jinzhi Pan, who instilled in me a great passion for science; and to my daughters, Melody and Maggie, of whom I am very proud.

Guanyu Wang
Department of Biology
South University of Science and Technology of China, Shenzhen
Department of Physics
George Washington University, Washington, DC

About the Author

Guanyu Wang obtained one PhD (in engineering) from Zhejiang University, China, in 1998 and one PhD (in medical sciences) from the University of Cologne, Germany, in 2005. As a postdoctoral fellow, he was initially at the School of Health Information of the University of Texas Health Science Center at Houston and then at the Department of Bioengineering, Rice University. He was an Assistant Professor of Pathology and Laboratory Medicine at the University of Texas Health Science Center at Houston and then Assistant Professor of Physics at George Washington University. He is now an Associate Professor of Biology at South University of Science and Technology of China, located in the city of Shenzhen.

Prelude: Yin-Yang Philosophy

In Chinese philosophy, the concept of Yin-Yang, literally meaning "shadow-light," is used to describe how polar opposites or seemingly contrary forces are interconnected and interdependent in the natural world. The concept lies at the heart of traditional Chinese philosophy and Chinese medicine, and is a central principle of different forms of Chinese martial arts and exercise. Yin and Yang interact with each other to form a mutual whole. This mutual whole, as well as the principle governing its interconversion, is known as Tao in Chinese philosophy. Everything has both Yin and Yang aspects and should thus be regarded as a mutual whole, as light could not be understood if darkness did not exist, and shadow cannot exist without light. This idea was documented in Chapter 42 of *Tao De Chin*, the most important Taoist classic text, by the famous sentence *Everything carries Yin yet embraces Yang, with a mellowing energy for harmony*. Many natural dualities (e.g., dark and light, female and male, low and high, cold and hot, water and fire, slowness and rapidness) are manifestations of Yin and Yang.

The reader will witness many Yin-Yang pairs in the book. In the global picture of diseases, diabetes and cancer constitute a Yin-Yang pair. Indeed, diabetes is characterized by insulin resistance (Yin), whereas certain cancers are characterized by insulin hypersensitivity (Yang). Normality is a single entity in the global picture, yet it also has Yin and Yang aspects. This book will demonstrate a key argument of traditional Chinese medicine, that a person's well-being depends on a well-balanced Yin and Yang. Even a single biomolecule is a Yin-Yang mutual whole. For example, a protein usually has both active (Yang) and inactive (Yin) states.

As a manifestation of Tao, Yin and Yang interact to convert into each other. In a cell, the switching between activation and inactivation of a biomolecule represents a typical example of Yin-Yang interconversion. The interconversion is especially important for the so-called *master regulator*, the hub of a network of molecular interactions. The protein kinase AKT (also known as protein kinase B) is such a master regulator, which plays a pivotal role in energy uptake and metabolism, cell growth, proliferation, and survival. It will turn out that one major function of the AKT pathway (the biomolecular network centering on AKT) is to realize and modulate the Yin-Yang interconversion, namely, the cycle between inactivated and activated AKT. It will turn out that normality is not a static balance between Yin and Yang, nor a gradual (graded) change from Yin to Yang and then back to Yin. It is the timely switching between full Yin and full Yang that renders normal physiology. The switching follows an interesting pattern known as bistability, which is characterized by all-or-none (full Yin and full Yang) and hysteresis (delayed switching in both directions). This bistable switch can be graphically represented, which can be regarded as an incarnation of Tao. The deformation of bistability would cause the blockage or retardation of Yin-Yang conversion, which underlies the progression of diseases such as cancer and diabetes.

Abbreviations

AMP	adenosine monophosphate
AMPK	AMP-activated protein kinase
ATP	adenosine triphosphate
ATPase	adenosine triphosphatase
BPG	bisphosphoglycerate
CoA	coenzyme A
DAG	diglyceride
DHAP	dihydroxyacetone phosphate
EGF	epidermal growth factor
F-1,6-BP	fructose 1,6-bisphosphate
FAD	flavin adenine dinucleotide
FFA	free fatty acid
GAP	glyceraldehyde 3-phosphate
GLUT	glucose transporter
GPCR	G-protein-coupled receptor
GSK	glycogen synthase kinase
HDL	high-density lipoprotein
IGF	insulin-like growth factor
IMTG	intramuscular triglyceride
IRS	insulin receptor substrate
LCFA	long-chain fatty acid
LDL	low-density lipoprotein
LPL	lipoprotein lipase
mTOR	mammalian target of rapamycin
NAD^+	nicotinamide adenine dinucleotide, oxidized form
NADH	nicotinamide adenine dinucleotide, reduced form
NADPH	nicotinamide adenine dinucleotide phosphate, reduced form
OMP	orotidine monophosphate
PDK1	3-phosphoinositide-dependent protein kinase-1
PdPC	phosphorylation and dephosphorylation cycle
PI3K	phosphoinositide 3-kinase
PKB	protein kinase B
PKC	protein kinase C
PPAR	peroxisome proliferator-activated receptor
PTM	posttranslational modification
RTK	receptor tyrosine kinase
TOR	target of rapamycin
TCA	tricarboxylic acid
TPI	triose phosphate isomerase
tRNA	transfer ribonucleic acid
TSC	tuberous sclerosis complex
UDP	uridine diphosphate
UMP	uridine monophosphate
UTP	uridine triphosphate
VLDL	very-low-density lipoprotein

Part I

Organismal Level: What Is Required

1 Food Intake and Energy Metabolism

Our health depends on well-balanced food intake and energy expenditure. Unlike plants, which can synthesize energy using carbon dioxide, water, and light (photosynthesis), humans must obtain basic organic molecules (glucose, amino acids, fat, vitamins, etc.) from food, which provides substrates for generating new tissue cells and energy to sustain our routine activities.

The digestion of these large molecules depends on the gastrointestinal tract, which is approximately nine meters long and extends from the mouth to the anus. Major organs of the gastrointestinal tract include the oral cavity, pharynx, esophagus, stomach, small intestine, and large intestine. Besides these major organs, digestion is also assisted by accessory organs such as the teeth, tongue, salivary glands, liver, gallbladder, and pancreas.

1.1 GASTROINTESTINAL TRACT

In the mouth, salivary glands secrete saliva, which naturally mixes with the food. Besides mucus and antimicrobial agents, saliva contains salivary amylase, an enzyme catalyzing the partial digestion of starch. Swallowing is rendered by wavelike contractions known as peristalsis, which involves the coordinated contraction of 25 pairs of muscles in the mouth, pharynx, larynx, and esophagus. In this way, the food is pushed all the way to the stomach.

1.1.1 Stomach

The stomach, located between the esophagus and the small intestine, is an important organ of the digestive tract in some animals, including vertebrates, echinoderms, insects, and molluscs. It is a muscular, hollow, dilated part of the digestion system. The stomach can store food, initiate digestion, kill bacteria (with the strong acidity of gastric juice), and move chyme (food processed by the stomach) into the small intestine. The mucosa of the stomach secretes pepsinogen (an inactive enzyme precursor) and hydrochloric acid. When the food reaches the lumen of the stomach, pepsinogen is converted into the active protein-digesting enzyme known as pepsin. The digestion and absorption capacity of the stomach is limited. The action of pepsin only contributes partially to the digestion of proteins and it does not contribute at all to the digestion of carbohydrates and fats. It should be noted that the digestion of starch begins in the mouth by the action of salivary amylase but amylase soon becomes inactivated by the strong acidity of gastric juice in the stomach. For absorption, the only commonly ingested substances that can be absorbed across the stomach wall are alcohol and aspirin.

1.1.2 Small Intestine

The small intestine, located between the stomach and the large intestine, is the longest part (approximately three meters) of the gastrointestinal tract. Its diameter is relatively small compared with the large intestine. The mucosa and submucosa of the small intestine form large folds, called plicae circulares, which can be observed with the eye. The surface area of the small intestine is further increased by microscopic folds of mucosa, called villi, and by folding of the apical plasma membrane of epithelial cells, called microvilli, which can be observed only with a microscope. The digestive enzymes of the small intestine are embedded within the plasma membrane of the microvilli. When eating begins, blood pours into villi and microvilli, further expanding their surface area and facilitating digestion.

Except for alcohol and aspirin, almost all the digestion products (carbohydrates, lipids, amino acids, calcium, iron, etc.) are absorbed through the small intestine. Indeed, the arrangement of villi and microvilli greatly increases the surface area for absorption and thus confers a rapid rate of absorption. In addition to providing a large surface area for absorption, the plasma membranes of the microvilli contain digestive enzymes that hydrolyze disaccharides, polypeptides, and other substrates. They are collectively called the brush border enzymes, because they remain attached to the cell membrane with their active sites exposed to the chyme, instead of being secreted into the lumen. Table 1.1 lists some brush border enzymes, together with their functions in digestion. A deficiency of these enzymes causes gastrointestinal disturbances and protein malnutrition. For example, lactase is a brush border enzyme that can digest milk sugar, or lactose. This enzyme is present in all children under the age of four but becomes inactive to some degree in most adults (especially Asians and Africans). Lactase deficiency can cause lactose intolerance, an unpleasant condition characterized by the accumulation of undigested lactose in the intestine, causing symptoms such as diarrhea, gas, and cramps. For these people, yogurt is a better diet choice than milk because it contains lactase (produced by the yogurt bacteria).

The small intestine and stomach are arguably the most resilient parts in the body. Certain animals such as snakes have such an extraordinary resilience that they can

TABLE 1.1
Brush Border Enzymes and Their Functions

Enzyme	Function
Sucrase	Digests sucrose to glucose and fructose
Maltase	Digests maltose to glucose
Lactase	Digests lactose to glucose and galactose
Aminopeptidase	Produces free amino acids, dipeptides, and tripeptides
Enterokinase	Required for activation of the protein-digesting enzyme trypsin, which enters the small intestine in pancreatic juice
Alkaline phosphatase	Removes phosphate groups from organic molecules
Ca^{2+}, Mg^{2+}-ATPase	Is needed for absorption of dietary calcium

even swallow animals larger than their own bodies. During the months between meals, a python's stomach and intestine atrophy. Once the snake begins to eat, however, it quickly revs up its digestive function. The first food to reach the gut — particularly proteins or their amino acids — stimulates a dramatic expansion of the gut lining. The stomach and intestine expand rapidly, whereby the absorptive surface area dramatically increases.

1.1.3 Large Intestine

The large intestine (colon) has little digestive function, but it does absorb water and electrolytes from the remaining chyme, as well as several B complex vitamins and vitamin K. The human colon contains many bacterial cells, whose number exceeds the number of cells of the body. In addition to producing vitamins B and K, these bacteria ferment (through anaerobic respiration) some indigestible molecules in the chyme and secreted mucus. These bacteria also produce short-chain fatty acids (less than five carbons long) to energize the colon epithelial cells and to aid the absorption of sodium, bicarbonate, calcium, magnesium, and iron in the large intestine. Finally, the large intestine passes waste products out of the body through the rectum and anal canal.

1.2 ACCESSORY DIGESTIVE ORGANS

1.2.1 Liver

The liver is the organ processing the blood and the products of digestion, before they enter the general circulation. The products of digestion are first absorbed into capillaries in the digestive tract, which converge into the hepatic portal vein. Then, the products of digestion pass through the hepatic portal vein and are absorbed into capillaries in the liver. The liver has many important functions. It detoxifies the blood by producing urea, uric acid, and other molecules that are less toxic than parent compounds, by chemically altering hormones and drugs, and by excreting some molecules in bile. It plays a pivotal role in energy balance. Depending on the body's energy status, the liver either removes glucose from the blood or secretes glucose into the blood. When the body has surplus sugar (e.g., after a meal), the liver can convert glucose into glycogen (glycogenesis) and triglycerides (lipogenesis) for storage. When the bloodstream is short of sugar (e.g., during fasting), the liver can generate glucose in two ways: either by breaking down the stored glycogen (glycogenolysis) or by converting nonsugar molecules into glucose (gluconeogenesis). If the body is still short of glucose, the liver also converts free fatty acids into ketone bodies (ketogenesis).

1.2.2 Pancreas

The pancreas has two major functions: exocrine and endocrine. The exocrine function involves the secretion of pancreatic juice through the pancreatic duct into the small intestine. Pancreatic juice contains water, bicarbonate, and a wide variety of enzymes. These enzymes, together with the brush border enzymes, act to digest food molecules in the small intestine. Among the enzymes in the pancreatic juice are trypsin, chymotrypsin, amylase (which digests starch to maltose and short chains of glucose molecules), elastase (which digests protein by cleaving internal peptide bonds), and lipase (which digests triglycerides by cleaving fatty acids from glycerol). The endocrine function of the pancreas is discussed in detail in Chapter 2. It involves the secretion of the peptide hormones insulin and glucagon into the bloodstream, by clusters of cells called the pancreatic islets, or islets of Langerhans.

1.3 GLUCOSE TRANSPORTATION

Every cell in the body needs energy to sustain its living status. The basic energy currency in the cell is adenosine triphosphate (ATP), which is used for three major purposes: the synthesis of proteins and other macromolecules, mechanical work in muscle contraction and cell migration, and ionic and neural pumping. ATP is largely derived from fuels such as glucose, amino acids, and fatty acids, which are the products of digestion and can reach tissue cells through the circulation. Through a series of biochemical reactions, the food molecules are catabolized so that ATP is generated. ATP is then utilized to meet the cells' energy demands.

The food molecules enter a cell primarily through facilitated diffusion, a spontaneous process (no energetic costs) aided by integral membrane proteins. Different kinds of food molecules have their corresponding kinds of transporter. That is, glucose is transported by glucose transporters and amino acids are transported by amino acid transporters. Because different tissues and organs have different metabolic profiles, there are many subtypes of transporter, even for transporting the same molecule. For example, glucose transporters (GLUTs) are a wide group of integral membrane proteins that facilitate the transportation of glucose through the plasma membrane. Currently, there are 14 types of GLUT known (GLUT1, GLUT2, . . . , GLUT14). The distribution of GLUTs is tissue specific. GLUT1 is a ubiquitous transporter that is expressed in virtually every cell type. GLUT3, the most efficient glucose transporter, is primarily expressed in neurons to meet the brain's absolute dependence on glucose. GLUT4 is expressed in adipose tissues and striated muscles, but not in other tissues.

GLUTs expose a single substrate-binding site toward either the outside or the inside of the cell. Binding of glucose to one site provokes a conformational change associated with transport. Once the GLUT reaches the other side of the plasma membrane, glucose is released. The process of glucose transportation can be regarded as an enzymatic reaction, which will be described in detail in Section 5.4, together with the associated Michaelis–Menten kinetics. In an enzymatic reaction, the biomolecule at the beginning of the process (known as the substrate) is converted into a different biomolecule (known as the product), and the process is catalyzed by a large biomolecule (known as the enzyme). Without the enzyme, the reaction can proceed, but at a much slower pace. The slowest known biological reaction takes about 1 trillion years to complete without the participation of enzymes, yet this reaction completes in only several milliseconds if the specific enzyme is added. In brief, an enzyme is responsible for the efficient conversion of the substrate into the product. As far as

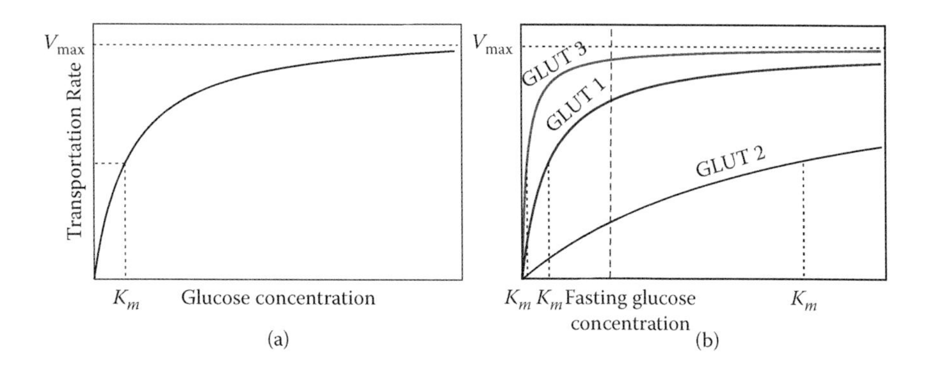

FIGURE 1.1 Michaelis–Menten kinetics of glucose transportation. (a) The rate of glucose transportation as a function of glucose concentration. (b) The kinetics of GLUT1, GLUT2, and GLUT3. Their Michaelis constants are also shown in their respective colors. The dash-dotted line indicates the fasting glucose concentration.

glucose transportation is concerned, the enzyme is GLUT, the substrate is glucose (at the loading site of the plasma membrane), and the product is also glucose (at the releasing site of the plasma membrane). Because we are primarily concerned with the transportation of glucose into a cell, the loading site usually refers to the blood and the releasing site usually refers to the cytoplasm. The reverse transportation is also common. During fasting, GLUT2 carries glucose out of the liver cells (hepatocytes) to replenish plasma glucose, in which case the loading site is the cytoplasm and the releasing site is the blood.

The kinetics of glucose transportation can be characterized by the parameters V_{max}, K_m, and k_{cat}/K_m (see Equation (5.2)). The parameter V_{max} represents the maximal transportation rate rendered by the GLUT. The parameter K_m, known as the Michaelis constant, is the half maximal effective concentration of the substrate, namely, the glucose concentration (of the loading site of the plasma membrane) at which the rate of transportation is half V_{max}. The parameter k_{cat} is the rate of production (the rapidness of glucose transportation) and k_{cat}/K_m measures the efficiency of GLUT. Figure 1.1(a) shows the kinetics of glucose transportation, from which the biological meanings of K_m and V_{max} can be easily seen.

The 14 types of GLUT can be divided into three classes. Class I comprises the well-characterized glucose transporters GLUT1, GLUT2, GLUT3, and GLUT4. Class II comprises GLUT5, GLUT7, GLUT9, and GLUT11. Class III comprises GLUT6, GLUT8, GLUT10, and GLUT12. In the following, class I GLUTs are described in detail.

- GLUT1 is a ubiquitous glucose transporter, responsible for the low level of basal glucose uptake required to sustain respiration in all cells. It is widely distributed in fetal tissues. In the adult, GLUT1 is expressed at its highest levels in erythrocytes and also in the endothelial cells of barrier tissues such as the blood–brain barrier. It is distributed asymmetrically between the luminal and abluminal membranes of the blood–brain barrier, with a ratio of roughly 1:4. The luminal membrane represents the key interface for glucose entry into the brain. Children with GLUT1 deficiency experience infantile seizures and developmental delay, caused by impaired glucose transportation across the blood–brain barrier. Expression levels of GLUT1 in cell membranes are increased by reduced glucose levels and decreased by increased glucose levels. GLUT1 has a very small Michaelis constant, $K_m \approx 1.5$ mM, far smaller than the plasma glucose concentration. Indeed, even during fasting, the plasma glucose concentration 4–6 mM is still threefold higher than K_m. Therefore, GLUT1 should be efficient in transporting glucose even when the plasma glucose concentration is minimal (at the fasting level). In Figure 1.1(b). One sees that the curve GLUT1 intersects the dash-dotted line (indicating the fasting glucose concentration) at a value of about $0.8V_{\text{max}}$. Therefore, the small K_m value of GLUT1 guarantees that every part of the body can receive the basal glucose input, because GLUT1 is expressed in almost every cell.

- GLUT2 is expressed in renal tubular cells, small intestinal epithelial cells, liver cells, and pancreatic beta cells. As a bidirectional transporter, it allows glucose to enter and leave the cell. Bidirectionality is very important. The liver cells need bidirectionality so that glucose can be absorbed when the plasma glucose concentration is high and released when the plasma glucose concentration is low. The pancreatic beta cells need a two-way flow of glucose so that the intracellular environment can accurately gauge the plasma glucose levels to determine the amount of insulin to be secreted. GLUT2 has a high capacity for glucose but low affinity, with $K_m = 15$–20 mM. Therefore, it loads glucose only when the glucose concentration is sufficiently high. The low affinity is necessary to avoid the simultaneous transportation of glucose in both directions, thus canceling out each other. If GLUT2 were of high affinity with glucose, then it would always load glucose and shuttle it in both directions, and the net flow of glucose would not be controlled by the glucose concentration. During fasting, the liver secretes glucose to the blood to nourish the whole body, in which event the reverse flow of glucose (from the blood to the liver) is undesired. Because the plasma glucose concentration is 4–6 mM in the fasting state, which is much smaller than the K_m of GLUT2, this reverse flow should be small. In Figure 1.1(b). The curve GLUT2 intersects the dash-dotted line at a small value, indicating a small reverse flow compared with the forward flow.
- GLUT3 is expressed in organs or tissues with high and rapid energy demands, such as the brain, sperm, embryonic tissues, white blood cells, and platelets. It is the major neuronal glucose transporter, present in both dendrites and axons of neurons. GLUT3 has a very small Michaelis constant K_m. There are great discrepancies about the exact value of K_m, probably due to its smallness and different kinds of glucose modification used in the experiments. Nevertheless, GLUT3 was consistently found to exhibit a lower K_m than the other class I transporters — even several times lower than that of GLUT1. Such a high affinity is crucial for the brain, which depends critically on glucose as

its sole energy source, to obtain glucose when the plasma glucose concentration is overly low. In Figure 1.1(b), one sees that GLUT3 is almost always saturated with glucose. In addition to high affinity, GLUT3 also has a high transportation capacity, because the turnover number k_{cat} of GLUT3 is the highest among all the GLUTs. GLUT3 has at least a fivefold greater transportation capacity than GLUT1 and GLUT4. This is particularly significant for the role of GLUT3 in the brain, as the ambient glucose levels surrounding the neurons are only 1–2 mM compared with 4–6 mM in serum. By using some experimental techniques, GLUT3 in rat cerebellar granule neurons was shown by Maher et al. [18] to have k_{cat} = 6500/s, which is significantly greater than k_{cat} = 1200/s (GLUT1 expressed in human erythrocytes, 3T3-L1 adipocytes, and oocytes) and k_{cat} = 1300/s (GLUT4 in 3T3-L1 adipocytes and oocytes). The high k_{cat}/K_m ratio implies that GLUT3 is a highly efficient transporter and explains why GLUT3 is the primary glucose transporter in the brain.

- GLUT4 is expressed in adipose tissues and striated muscles (skeletal and cardiac). It is responsible for insulin-mediated glucose translocation into the cell. When the plasma insulin concentration is low, GLUT4 is sequestered in intracellular vesicles in muscle and fat cells. When the plasma insulin concentration becomes sufficiently high, the vesicles translocate to the plasma membrane and fuse with it, whereby GLUT4 transporters are inserted onto the plasma membrane and become available for transporting glucose. The process is controlled by the PI3K-AKT-TOR signaling pathway, which is described in detail in Chapter 7. GLUT4 has a Michaelis constant K_m about 5 mM, which appears to be larger than that of GLUT3 and smaller than that of GLUT2. Since its discovery in the late 1980s, GLUT4 has received more attention than any other membrane transporter, probably due to its apparent importance in whole-body glucose homeostasis, the intricate mechanism of its control by insulin, and the consequences of disruption of the control mechanism (causing prevalent insulin-resistant states such as obesity and type 2 diabetes).

1.4 METABOLIC PROFILE OF ORGANS

The diversity of glucose transporters has already implied that different organs differ greatly in utilizing glucose. Because fatty acid is another key fuel besides glucose, the heterogeneity of energy metabolism among organs may be even more dramatic. In the following, energy metabolism of many organs is described, including the brain, muscle, liver, kidney, and adipose tissue. It turns out that each organ has a unique metabolic profile.

1.4.1 BRAIN

The human brain has the same general structure as those of other mammals. But it is much larger than that of any other mammal in relation to body size. The adult human brain weighs on average about 1.5 kg, with a volume of about 1130 cm^3 in women and 1260 cm^3 in men, although there is substantial individual variation. Large animals such as whales and elephants have larger brains in absolute terms. But when measured in terms of the encephalization quotient, which compensates for body size, the human brain is almost twice as large as the brain of the bottlenose dolphin and three times as large as the brain of a chimpanzee.

Unlike most other organs that can store energy (in the form of glycogen or triglyceride), the brain lacks fuel stores and hence requires a continuous supply of energy. Because fatty acids cannot cross the blood–brain barrier, the brain depends absolutely on glucose as its sole energy source. Every day the brain consumes about 120 g of glucose, which corresponds to an energy value of about 1760 kJ (420 kcal). This accounts for some 60% of glucose utilization by the whole body in the resting state. Even when the total energy (including fatty acids) is considered, the brain still consumes about 20% of total energy, which is more than any other organ. Most of the energy is used for neural pumping, namely, the maintenance of the Na^+–K^+ membrane potential that is necessary for transmitting nerve impulses. Neurons primarily use GLUT3 to transport glucose. Although GLUT3 was found to be expressed in various tissues, it is most specifically expressed in neurons, found predominantly in axons and dendrites and also, but less prominently, in the cell body. During prolonged starvation, ketone bodies generated by the liver partly replace glucose as fuel for the brain.

1.4.2 Skeletal Muscle

Skeletal muscle exists throughout the human body and is one of three major muscle types, the others being cardiac and smooth muscle. Skeletal and cardiac muscles are collectively known as striated muscle, because they have striations, or stripes, that extend across the width of the muscle cell. As their name suggests, most skeletal muscles are attached to bones by bundles of collagen fibers known as tendons. Skeletal muscles are made up of individual components known as myocytes, or "muscle cells," sometimes colloquially called "muscle fibers."

Skeletal muscles use glucose, lipids, and ketone bodies as their fuel. It was once suggested that glucose was the dominating energy substrate and that lipids were not directly oxidized but first converted into glucose and glycogen in the liver. Today, it is well known that lipids are important fuels for skeletal muscles. Especially in the resting state, skeletal muscles prefer to use lipids as their major fuel to meet 85% of their energy requirements. Lipids as fuel for energy provision originate from several different sources: albumin-bound long-chain fatty acids (LCFA) in the blood, very-low-density lipoprotein-triacylglycerols (VLDL-TG), fatty acids from triacylglycerol located in the muscle cells, and possibly fatty acids liberated from adipose tissues adhering to the muscle cells.

During exercise, as activity becomes more intense, more and more glycogen and blood glucose are used as energy sources. Skeletal muscles store about 5000 kJ (1200 kcal) of glycogen, which is about three-fourths of all the glycogen in the body. When needed, glycogen is readily converted into glucose 6-phosphatase. Skeletal muscles break down glucose anaerobically (glycolysis) for the first 45 to 90 seconds of moderate-to-heavy exercise because the cardiopulmonary system requires that amount of time to sufficiently increase the oxygen supply to the exercising muscles. If exercise is moderate, aerobic respiration contributes the major portion of the skeletal muscle energy requirements following the first 2 minutes of exercise. During anaerobic respiration, glucose is converted into pyruvate and then into lactate, which flows to the liver to be converted into glucose again (the Cori cycle). The transamination of pyruvate in the skeletal muscle produces alanine, which also flows to the liver. The liver absorbs the alanine, removes the nitrogen for disposal as urea, and processes the pyruvate to glucose or fatty acids. These interchanges shift part of the metabolic burden of muscle to the liver.

When the plasma glucose and insulin concentrations are small (e.g., in the fasting state), skeletal muscles may use GLUT1 to acquire a low-level, basal glucose supply. However, massive glucose uptake does not happen in this state, because GLUT4 does not translocate to the plasma membrane of the cell due to insufficient insulin stimulation. When the plasma insulin concentration becomes high, GLUT4 begins to transport glucose into skeletal muscles. Besides insulin stimulation, muscle contraction (e.g., during exercise) also signals the translocation of GLUT4 to the plasma membrane, enabling glucose entry into the contracting muscles.

1.4.3 Cardiac Muscle

Cardiac muscle is a type of striated muscle found in the heart. Cardiomyocytes, the cells that constitute cardiac muscles, usually contain one or two nuclei. In very rare cases, cardiomyocytes can contain three or four nuclei. Coordinated contractions of cardiomyocytes in the heart propel blood out of the atria and ventricles to the circulatory system, rendering the systole of the heart. Like all tissues in the body, cardiomyocytes rely on an ample blood supply to deliver oxygen and nutrients and to remove waste products such as carbon dioxide. In accordance with the ample blood supply, cardiomyocytes have numerous myoglobins (oxygen-storing pigment) and a large number of mitochondria, enabling continuous aerobic respiration via oxidative phosphorylation.

The heart is so adapted to aerobic metabolism that it is unable to pump sufficiently in ischemic conditions. At basal metabolic rates, about 1% of energy is derived from anaerobic metabolism. In situations of metabolic stress, anaerobic metabolism is increased (through utilizing more glucose so that more glycolysis occurs) to counteract energy depletion. Anaerobic metabolism can increase to 10% under moderately hypoxic conditions. Under more severe hypoxic conditions, however, the energy liberated from glycolysis and lactate production is insufficient to sustain ventricular contractions.

Under basal aerobic conditions, 60% of energy comes from fat (free fatty acids and triglycerides), 35% from carbohydrates (primarily as glucose), and 5% from amino acids and ketone bodies. However, these proportions vary widely according to nutritional state. For example, during starvation, lactate can be recycled by the heart, which can then be burned aerobically in the citric acid cycle, liberating much more ATP (see the next section). In the condition of diabetes, more fat and less carbohydrate is used due to the reduced induction of GLUT4 to the plasma membrane of the cell.

The contraction of skeletal and cardiac muscles plays a role in bringing GLUT4 transporters to the cell surface. This is particularly relevant to cardiac muscle due to its spontaneous and continuous contractions.

1.4.4 Smooth Muscle

Unlike skeletal and cardiac muscles, smooth muscles are not striated. They are found within the walls of blood vessels such as in the tunica media layer of large (aorta) and small arteries, arterioles, and veins. These smooth muscles are termed vascular smooth muscles. They are also found in lymphatic vessels, the urinary bladder, uterus, male and female reproductive tracts, gastrointestinal tract, respiratory tract, arrector pili of skin, the ciliary muscle, and iris of the eye. In addition, the glomeruli of the kidneys contain mesangial cells which are similar to smooth muscle cells. The structure and function of smooth muscle cells in different organs are basically the same. However, the inducing stimuli differ substantially among these smooth muscles, because they realize different functions in different organs.

Compared with striated muscles, smooth muscle's metabolism is slow and generates a low concentration of metabolites. It was found that smooth muscle metabolism is compartmented such that mitochondrial function fuels contraction and that much glycolytic ATP production is used for membrane pumps. Even in the absence of mitochondrial function (cyanide inhibition), smooth muscle metabolism is able to maintain ATP concentration. Therefore, the vessels are able to adapt to stressful metabolic conditions.

1.4.5 Liver

The liver, constituting 2% to 4% of body weight, is a master regulator of the organism's metabolism. As the portal of entry into the blood for food molecules absorbed by the intestine, the liver is able to regulate the level of many metabolites in the blood. It is essential for providing fuel to the brain, especially in the fasting state. Glucose metabolism is a good example to explain the liver's function. After a meal, only about one third of the glucose is used directly by tissue cells, while the remaining glucose enters the liver. Glucose absorbed into the liver is converted into glucose 6-phosphate by hexokinase and liver-specific glucokinase. While the liver can certainly use glucose 6-phosphate to meet its own energy needs, most of it is converted into glycogen for storage (as much as 1700 kJ). If still in excess, glucose 6-phosphate is metabolized to acetyl coenzyme A (CoA), which is used to form fatty acids, cholesterol, and bile salts. These reductive biosyntheses need NADPH, which is supplied by glucose 6-phosphate through the pentose phosphate pathway. During fasting, glycogen stored in the liver turns into glucose to be released into the blood. The liver can also produce glucose by carrying out gluconeogenesis. The main precursors for gluconeogenesis are lactate and alanine from muscle, glycerol from adipose tissue, and glucogenic amino acids from the diet.

The liver is also important in regulating lipid metabolism. The liver can either promote the conversion of fatty acids into triglycerides for storage in adipose tissues or promote the conversion of fatty acids into ketone bodies for utilization by tissue cells. The fate of fatty acids is determined by the level of malonyl CoA. When fuels are abundant, the level of malonyl CoA is high, which inhibits carnitine acyltransferase I. Consequently, long-chain fatty acids cannot be esterified into carnitine and traverse the inner mitochondrial membrane, which prevents β oxidation and the formation of ketone bodies. Therefore, fatty acids are absorbed by adipose tissues and are incorporated into triglycerides. When fuels are scarce, the level of malonyl CoA is low, which allows fatty acids to enter the mitochondrial matrix of tissue cells for conversion into ketone bodies.

Like glucose metabolism, the majority of dietary amino acids enter the liver and only some remain in the blood for peripheral utilization. Because tRNAs (which participate in protein synthesis) have a higher affinity with amino acids than the enzymes catalyzing amino acid catabolism, amino acids are primarily used for anabolism rather than catabolism. Under conditions such as starvation, amino acid catabolism may occur. The first step is the removal of nitrogen, which is subsequently processed to urea. The degradation of amino acids then produces α-ketoacids, which are used for gluconeogenesis, fatty acid synthesis, and the liver's own fuel. By using α-ketoacids, the liver spares the fuels it exports to other tissue cells. Indeed, the main role of glycolysis in the liver is to form building blocks for biosyntheses. The liver cannot use acetoacetate as a fuel, because it has little of the transferase needed for acetoacetate's activation to acetyl CoA.

1.4.6 KIDNEY

Located at the rear of the abdominal cavity in the retroperitoneum, the kidneys receive blood from the paired renal arteries, which drain into the paired renal veins. The kidney is essential in the urinary system and also serves homeostatic functions such as the regulation of electrolytes, maintenance of acid–base balance, and regulation of blood pressure (via maintaining salt and water balance). It is a natural filter of the blood, and removes wastes which are diverted to the urinary bladder. In producing urine, the kidney excretes wastes such as urea and ammonium, and it is also responsible for the reabsorption of water, glucose, and amino acids. The kidney also produces hormones, including calcitriol, erythropoietin, and the enzyme renin.

The blood plasma is filtered every 25 min or so in the kidney, which produces 1 to 2 liters of urine every day. Water and water-soluble materials (e.g., glucose) are reabsorbed to prevent wasteful loss. The reabsorbed glucose is carried into the kidney cells by the sodium-glucose cotransporter, powered by the Na^+–K^+ gradient (which is maintained by Na^+–K^+ ATPase). The reabsorption costs a lot of energy, which explains why the kidney consumes 10% of the oxygen while it only constitutes 0.5% of body mass. During starvation, the kidney becomes an important site of gluconeogenesis and may contribute as much as half of the plasma glucose.

1.4.7 ADIPOSE TISSUE

Adipose tissues constitute an enormous reservoir of metabolic fuel in the body. The mass of triglyceride is typically 15 kg for man, which is equivalent to 565,000 kJ (135,000 kcal) of energy.

After a meal, dietary fat is broken down into free fatty acids (FFAs) and monoglycerides by pancreatic lipase, in order to be absorbed by the intestine. Once across the intestinal barrier, they are reformed into triglycerides and packaged into chylomicrons or liposomes, which are directly absorbed by adipose tissues or processed by the liver first. On one hand, lipoprotein lipase (LPL) in the capillaries of adipose tissues hydrolyses triglycerides in the chylomicrons or liposomes. Fatty acids thus released are taken up into the adipocytes for storage. On the other hand, the liver processes chylomicron remnants and liposomes into various lipoprotein forms, in particular very-low-density lipoproteins (VLDLs), which are then transported to the blood. In peripheral tissues, LPLs digest part of the VLDL (by removing triglycerides) into low-density lipoprotein (LDL) and FFAs. The VLDL remnants may also become LDLs by absorbing cholesterol from other circulating lipoproteins. Another type of lipoprotein known as high-density lipoprotein (HDL) collects cholesterol, glycerol, and fatty acids from the blood and transports them to the liver for breakdown and excretion. LDLs enter adipocytes through LDL receptors and are then converted into free fatty acids, cholesterol, and other components. The principal task of adipose tissue is to activate these fatty acids and transfer the resulting CoA derivatives to glycerol in the form of glycerol 3-phosphate. This essential intermediate in lipid biosynthesis comes from the reduction of the glycolytic intermediate dihydroxyacetone phosphate. Therefore, adipocytes need glucose for synthesizing triglyceride. Not all FFAs are deposited. There is always a proportion of FFAs that joins the plasma FFA pool in a process sometimes called spillover. Spillover fatty acids may constitute 40% to 50% of the total plasma FFA pool in the postprandial period.

In the fasting state, triglycerides are converted into fatty acids and glycerol by an intracellular lipase, which is catalyzed by the hormone epinephrine. The hormone stimulates the formation of cyclic AMP, the intracellular messenger in the amplifying cascade, which activates the lipase. The amplifying cascade is important in the release of the first fatty acid from a triglyceride, which is the rate-limiting step. Triglycerides in adipocytes are continuously being hydrolyzed and resynthesized. Glycerol derived from their hydrolysis is exported to the liver. Most of the fatty acids resulting from hydrolysis are reesterified if glycerol 3-phosphate is abundant. In contrast, they are released into the plasma if glycerol 3-phosphate is scarce because of a paucity of glucose. The glucose level within adipocytes is thus a major factor in determining whether or not fatty acids are released into the plasma. In the blood, free fatty acids are bound by serum albumin and are transported to the tissue needing fuel. Once the fatty acids reach the target tissue, they are released by serum albumin and diffusion into the cell. In the cell, fatty acids are broken down in mitochondria and/or peroxisomes to generate acetyl-CoA, the entry molecule for the citric acid cycle, through which ATP is generated to sustain the cell's needs.

1.5 GLUCOSE METABOLISM

Among all the metabolic fuels (ketone bodies, glucose, fatty acids, and amino acids), glucose is the most commonly used one and is especially important. Under nonstarvation conditions, glucose is the only fuel used by the brain of a mammal. Besides, glucose is the only fuel that erythrocytes can use. The next chapter will be devoted to the study of whole-body glucose homeostasis, namely, the intricate interplay between glucose and insulin that results in fairly constant plasma glucose and insulin concentrations even in the face of frequent meal perturbations. This section focuses on events after the entry of glucose into the cell: how is it metabolized into ATP for energy utilization? How does it provide substrates for the cell to synthesize macromolecules?

1.5.1 Glycolysis

Glycolysis, as the first step of glucose metabolism, takes place in the cytoplasm. It consists of a sequence of biochemical reactions that metabolizes one glucose molecule to two molecules of pyruvate with the concomitant net production of two molecules of ATP. Pyruvate can be further metabolized to lactate (lactic acid fermentation) or ethanol (alcoholic fermentation). Glycolysis is an anaerobic process which does not require the participation of O_2.

The first stage of glycolysis involves the conversion of glucose into fructose 1,6-bisphosphate (F-1,6-BP). After entering the cell, glucose is phosphorylated at the sixth carbon to form glucose 6-phosphate, which consumes one ATP:

$$\text{Glucose} + \text{ATP} \rightarrow \text{Glucose 6-phosphate} + \text{ADP} + \text{H}^+. \tag{1.1}$$

The phosphorylation is catalyzed by the enzyme hexokinase in most cells, and by the enzyme glucokinase in certain cells (most notably liver cells) of higher animals. Not being a substrate of GLUTs, glucose 6-phosphate is trapped within the cell. The next step is the isomerization of glucose 6-phosphate to fructose 6-phosphate. The reaction is catalyzed by phosphoglucose isomerase, which opens the six-membered ring of glucose 6-phosphate, catalyzes the isomerization, and then promotes the formation of the five-membered ring of fructose 6-phosphate:

$$\text{Glucose 6-phosphate} \leftrightharpoons \text{Fructose 6-phosphate}. \tag{1.2}$$

Then, fructose 6-phosphate is phosphorylated at the first carbon to form fructose-1,6-BP, which consumes one ATP:

$$\text{Fructose 6-phosphate} + \text{ATP} \rightarrow \text{F-1,6-BP} + \text{ADP} + \text{H}^+. \tag{1.3}$$

The phosphorylation is catalyzed by the enzyme phosphofructokinase. Note that the two monophosphoryl groups are present separately. That is why the prefix *bis-* is used, instead of the prefix *di-*, which signifies two connected phosphorl groups (e.g., adenosine diphosphate).

The second stage involves the splitting of F-1,6-BP into two three-carbon fragments: glyceraldehyde 3-phosphate (GAP) and dihydroxyacetone phosphate (DHAP), catalyzed by the enzyme aldolase. This reaction is reversible, which means that GAP and DHAP can form F-1,6-BP:

$$\text{F-1,6-BP} \leftrightharpoons \text{GAP} + \text{DHAP}. \tag{1.4}$$

DHAP and GAP can convert into each other, with catalyzation by triose phosphate isomerase (TPI):

$$\text{DHAP} \leftrightharpoons \text{GAP}. \tag{1.5}$$

The interconversion is important, because only GAP is on the direct pathway of glycolysis. That is, DHAP has to be converted into GAP in order to follow the remaining steps of glycolysis. At equilibrium, DHAP (GAP) represents 96% (4%) of the triose phosphate. Nevertheless, DHAP can readily convert into GAP because the subsequent reactions of glycolysis remove GAP rapidly.

The third stage involves the oxidation of the three-carbon fragments to pyruvate. First, GAP converts into 1,3-bisphosphoglycerate (1,3-BPG), a reaction catalyzed by glyceraldehyde 3-phosphate dehydrogenase. Concomitantly, nicotinamide adenine dinucleotide (NAD^+) is reduced to NADH:

$$\text{GAP} + \text{P}_\text{i} + \text{NAD}^+ \leftrightharpoons \text{1,3-BPG} + \text{NADH} + \text{H}^+. \tag{1.6}$$

1,3-BPG is an energy-rich molecule with a greater phosphoryl-transfer potential than that of ATP. Therefore, 1,3-BPG can recharge ADP to form ATP, with 3-phosphoglycerate as another product. The process is catalyzed by phosphoglycerate kinase:

$$\text{1,3-BPG} + \text{ADP} \leftrightharpoons \text{3-Phosphoglycerate} + \text{ATP}. \tag{1.7}$$

3-Phosphoglycerate converts into 2-phosphoglycerate, with the position of the phosphoryl group shifted. The reaction is catalyzed by phosphoglycerate mutase:

$$\text{3-Phosphoglycerate} \leftrightharpoons \text{2-Phosphoglycerate}. \tag{1.8}$$

The dehydration of 2-phosphoglycerate results in the production of phosphoenolpyruvate, which has a high phosphoryl-transfer potential:

$$\text{2-Phosphoglycerate} \leftrightharpoons \text{Phosphoenolpyruvate} + \text{H}_2\text{O}. \tag{1.9}$$

Phosphoenolpyruvate transfers a phosphoryl group to ADP to form ATP and pyruvate. This reaction, catalyzed by pyruvate kinase, is virtually irreversible:

$$\text{Phosphoenolpyruvate} + \text{ADP} + \text{H}^+ \rightarrow \text{Pyruvate} + \text{ATP}. \tag{1.10}$$

Taking the three stages together, the net reaction of glycolysis can be represented by

$$\begin{aligned} &\text{Glucose} + 2\text{P}_\text{i} + 2\text{ADP} + 2\text{NAD}^+ \rightarrow \\ &2\text{Pyruvate} + 2\text{ATP} + 2\text{NADH} + 2\text{H}^+ + 2\text{H}_2\text{O}. \end{aligned} \tag{1.11}$$

Note that two molecules of ATP are generated during glycolysis.

The reaction reduces NAD^+ to NADH. Because the amount of NAD^+ is very limited in the cell, glycolysis cannot proceed for long if NAD^+ is not regenerated. Therefore, the final process in the pathway is the regeneration of NAD^+ through the metabolism of pyruvate. While the reaction from glucose to pyruvate is almost the same for most organisms, the fate of pyruvate is different. It is converted into lactic acid in some organisms and into ethanol in others. Carbon dioxide is also a derivative of pyruvate.

1.5.2 Alcoholic Fermentation

In yeast and some other microorganisms, pyruvate is converted into ethanol as the outcome of glycolysis. Pyruvate is first converted into acetaldehyde, catalyzed by pyruvate decarboxylase. Acetaldehyde is then reduced to ethanol by NADH, whereby NAD^+ is regenerated. The net reaction (from glucose to ethanol) can be represented by:

$$\begin{aligned}&\text{Glucose} + 2\text{P}_\text{i} + 2\text{ADP} \rightarrow \\ &2\text{Ethanol} + 2\text{CO}_2 + 2\text{H}_2\text{O} + 2\text{ATP}.\end{aligned} \tag{1.12}$$

NAD^+ and NADH, although crucial for the overall process, do not appear in this reaction. Therefore, there is no net oxidation–reduction in the conversion of glucose into ethanol. The regeneration of NAD^+ during the process of reducing pyruvate to ethanol (or lactate, see below) sustains the continued operation of glycolysis under anaerobic conditions.

1.5.3 Lactic Acid Fermentation

In some microorganisms, pyruvate is converted into lactate as the outcome of glycolysis. The reaction also takes place in the cells of higher organisms (e.g., humans) when O_2 becomes limited. During intense contraction, for example, muscles often need to use lactic acid fermentation to generate ATP. The reaction is simpler than alcoholic fermentation: pyruvate is directly reduced to lactate by NADH, whereby NAD^+ is regenerated. The net reaction (from glucose to lactate) can be represented by:

$$\text{Glucose} + 2\text{P}_\text{i} + 2\text{ADP} + 4\text{H}^+ \rightarrow 2\text{Ethanol} + 2\text{CO}_2 + 2\text{H}_2\text{O} + 2\text{ATP}, \tag{1.13}$$

which has no net oxidation–reduction. Lactic acid fermentation is especially important for cancer cells. The growth of a tumor is often associated with the hypoxia condition, because the normal oxygen supply from blood vessels is insufficient for the aerobic respiration of cancer cells that are rapidly dividing. The limitation of aerobic respiration makes glycolysis and lactic acid fermentation the primary source of ATP generation. Because glycolysis is very inefficient (two ATP yields per glucose molecule) compared with oxidative phosphorylation (which yields more than 30 ATPs per glucose molecule), cancer cells compensate by consuming large amounts of glucose, much more than glucose consumed by normal cells. Cancer cells obtain this ability through constitutive activation of some oncoproteins (see Chapters 7 and 9), which increases the expression of many glycolytic enzymes and glucose transporters.

1.5.4 Citric Acid Cycle

We have discussed two outcomes of pyruvate: ethanol and lactic acid. Then, how does pyruvate metabolize to CO_2 as the third outcome? This metabolism depends on the citric acid cycle, also known as the tricarboxylic acid (TCA) cycle or the Krebs cycle—a series of chemical reactions used by all aerobic organisms to generate energy through the oxidization of acetyl CoA derived from glucose, fats, and amino acids into carbon dioxide. In addition, the cycle provides precursors, including certain amino acids, as well as the reducing agent NADH that is used in numerous biochemical reactions. In eukaryotes, the reactions of the citric acid cycle take place inside mitochondria.

In glucose catabolism, the precursor of acetyl CoA is pyruvate. Under aerobic conditions, pyruvate is transported into mitochondria in exchange for OH^- by the pyruvate carrier. In the mitochondrial matrix, pyruvate is oxidatively decarboxylated by the pyruvate dehydrogenase complex to form acetyl CoA:

$$\text{Glucose} + 2\text{P}_\text{i} + 2\text{ADP} + 2\text{H}^+ \rightarrow 2\text{Lactate} + 2\text{H}_2\text{O} + 2\text{ATP}. \quad (1.14)$$

The citric acid cycle begins with the condensation of a four-carbon unit, oxaloacetate, and a two-carbon unit, the acetyl group of acetyl CoA. Oxaloacetate reacts with acetyl CoA and H_2O to yield citrate and CoA:

$$\text{Acetyl CoA} + \text{Oxaloacetate} + \text{H}_2\text{O} \rightarrow \text{Citrate} + \text{H}^+ + \text{CoA}. \quad (1.15)$$

The original citrate molecule is not very well suited for oxidative decarboxylation, because the tertiary hydroxyl group is not properly located. Thus, citrate has to be isomerized into isocitrate to enable the six-carbon unit to undergo oxidative decarboxylation. The isomerization is accomplished by dehydration and then hydration:

$$\text{Citrate} \leftrightharpoons cis\text{-Aconitate} + \text{H}_2\text{O}, \quad (1.16)$$

$$cis\text{-Aconitate} + \text{H}_2\text{O} \leftrightharpoons \text{Isocitrate}. \quad (1.17)$$

Isocitrate then undergoes oxidative decarboxylation (the first oxidation–reduction reaction in the citric acid cycle). The reaction leads to oxalosuccinate, an unstable β-ketoacid. Upon binding with the enzyme isocitrate dehydrogenase, it loses CO_2 to form α-ketoglutarate. The reaction also generates, for the first time in the citric acid cycle, NADH, which is a high-transfer-potential electron carrier:

$$\text{Isocitrate} + \text{NAD}^+ \leftrightharpoons \text{CO}_2 + \text{NADH} + \alpha\text{-Ketoglutarate}. \quad (1.18)$$

It should be noted that the rate of formation of α-ketoglutarate is important in determining the overall rate of the citric acid cycle. Then, the second oxidation–reduction reaction proceeds, whereby succinyl CoA is formed from α-ketoglutarate:

$$\alpha\text{-Ketoglutarate} + \text{NAD}^+ + \text{CoA} \leftrightharpoons \text{CO}_2 + \text{NADH} + \text{Succinyl CoA}. \quad (1.19)$$

Note the close resemblance between reaction (1.19) and reaction (1.14), which is not strange because pyruvate is an α-ketoacid. Succinyl CoA is energy-rich. The cleavage of the thioester bond of succinyl CoA releases energy that is sufficient for the phosphorylation of a purine nucleoside diphosphate, usually GDP:

$$\text{Succinyl-CoA} + \text{P}_\text{i} + \text{GDP} \leftrightharpoons \text{Succinate} + \text{GTP} + \text{CoA}. \quad (1.20)$$

Succinate is oxidized to fumarate by the enzyme succinate dehydrogenase. In this reaction, FAD is the hydrogen acceptor, because the free energy change is insufficient to reduce NAD^+. FAD is nearly always the electron acceptor in oxidations that remove two hydrogen atoms from a substrate. Note that FAD is covalently attached to the enzyme during the reaction, and the compound is denoted by E-FAD:

$$\text{Succinate} + \text{E-FAD} \rightleftharpoons \text{Fumarate} + \text{E-FADH}_2. \tag{1.21}$$

The next step is the hydration of fumarate to form L-malate:

$$\text{Fumarate} + H_2O \rightleftharpoons \text{L-Malate}. \tag{1.22}$$

Finally, malate is oxidized to form oxaloacetate. This reaction is catalyzed by malate dehydrogenase, and NAD is again the hydrogen acceptor:

$$\text{L-Malate} + NAD^+ \rightleftharpoons \text{Oxaloacetate} + NADH + H^+. \tag{1.23}$$

In summary, the net reaction of the citric acid cycle can be represented by

$$\begin{aligned} &\text{Acetyl CoA} + P_i + FAD + GDP + 3NAD^+ + 2H_2O \rightarrow \\ &2CO_2 + FADH_2 + 3NADH + GTP + 2H^+ + CoA. \end{aligned} \tag{1.24}$$

During one citric acid cycle, an acetyl group is oxidized to generate two molecules of carbon dioxide, through a series of oxidation–reduction reactions. Although intermediate reactants contain multiple carbons (oxaloacetate: four-carbon; citrate: six-carbon; α-ketoglutarate: five-carbon; succinate: four-carbon), only two carbon atoms enter the cycle as an acetyl unit and two carbon atoms leave the cycle (in the form of two CO_2). This explains why fat cannot convert into glucose for utilization. Because oxaloacetate is the precursor of glucose, one may hope that carbon atoms from fat can be transferred to glucose via oxaloacetate, an intermediate of the citric acid cycle. However, although two carbon atoms from fat enter the cycle as acetyl CoA, two carbon atoms are lost as CO_2 before oxaloacetate is formed. Thus, there is no net synthesis of glucose from fats. In the citric acid cycle, three hydride ions (H^-, six electrons) are transferred to three NAD^+ molecules, whereas one pair of hydrogen atoms (hence, two electrons) is transferred to one FAD molecule. Therefore, the essence of the citric acid cycle is to harvest high-energy electrons from carbon fuels and store them in NADH and $FADH_2$ (for use later).

1.5.5 Oxidative Phosphorylation

Humans need plenty of ATP to meet their basal demands. A sedentary man of 145 lbs requires about 8000 kJ energy for a day's activity. This is a big energy budget, equivalent to about 80 kg of ATP. However, a regular man possesses only about 0.25 kg of ATP at any one time. This problem is solved by recycling ADP back to ATP frequently, at a rate about 300 times per day. This recycling takes place primarily through oxidative phosphorylation. In eukaryotes, oxidative phosphorylation takes place in mitochondria.

As already mentioned, the citric acid cycle does not directly generate ATP. It only stores high-energy electrons in NADH and $FADH_2$, which have a high potential of transferring the electrons to O_2. To reach O_2, the electrons must flow through a chain of three large protein complexes located in the mitochondrial inner membrane: NADH-Q oxidoreductase (Complex I), Q-cytochrome c oxidoreductase (Complex III), and cytochrome c oxidase (Complex IV). These large transmembrane complexes contain multiple oxidation–reduction centers, including quinones, flavins, iron-sulfur clusters, hemes, and copper ions. Electron flow through the electron-transport chain leads to the transport of protons across the inner mitochondrial membrane. A fourth large protein complex, called succinate-Q reductase (Complex II), contains the succinate dehydrogenase that generates $FADH_2$ in the citric acid cycle. Electrons from this $FADH_2$ enter the electron-transport chain at Q-cytochrome oxidoreductase. Complex II, in contrast with the other complexes, does not pump protons. Complexes I, II, and III appear to be associated with a supramolecular complex known as the respirasome. Such supramolecular complexes facilitate the rapid transfer of substrate and prevent the release of reaction intermediates. The flow of electrons from NADH to O_2 can be represented by

$$\mathrm{NADH} + 0.5\, \mathrm{O_2} + \mathrm{H^+} \leftrightharpoons \mathrm{H_2O} + \mathrm{NAD^+}, \tag{1.25}$$

which is an exergonic process, with a net change of Gibbs free energy -220.1 kJ/mol. This energy release is sufficient for the synthesis of ATP, which is an endergonic process, with a net change of Gibbs free energy $+30.5$ kJ/mol.

$$\mathrm{ADP} + \mathrm{P_i} + \mathrm{H^+} \leftrightharpoons \mathrm{H_2O} + \mathrm{ATP}. \tag{1.26}$$

How does the oxidation of NADH render ATP synthesis? It is realized by a proton gradient across the inner mitochondrial membrane that couples electron transport and ADP phosphorylation. As electrons passes through the electron-transport chain, protons are pumped from the matrix of the mitochondrion to the cytoplasmic side of the inner mitochondrial membrane. An electric field is thus generated, with the matrix side negative (because the matrix has a lower concentration of H^+). This unequal distribution of protons is called the proton-motive force, which stores energy. It is characterized by two gradients. The chemical (proton) gradient is characterized by a lower pH value outside than inside, with 1.4 unit difference. The charge gradient is characterized by the 0.14 V membrane potential, with the outside being positive. This potential corresponds to a free energy of 21.8 kJ per mole of protons. To equalize the distribution, protons flow back into the mitochondrial matrix, which drives the synthesis of ATP by ATP synthase, a molecular assembly in the inner mitochondrial membrane, also known as Complex V. In this way, the oxidation of fuels and the phosphorylation of ADP are coupled by a proton gradient across the inner mitochondrial membrane.

ATP synthase catalyzes the formation of ATP from ADP and orthophosphate.

$$\mathrm{ADP^{3-}} + \mathrm{HPO_4^{2-}} + \mathrm{H^+} \leftrightharpoons \mathrm{H_2O} + \mathrm{ATP^{4-}}. \tag{1.27}$$

ATP synthase is a large, complex enzyme that looks like a ball on a stick, with the stick (F_0 subunit) embedded in the inner mitochondrial membrane and the ball (F_1 subunit) protruded into the mitochondrial matrix. The F_1 subunit consists of five types of polypeptide chains, α, β, γ, δ, and ε. One ATP synthase includes three α and three

β subunits, which are arranged alternately in a hexametric ring. The α subunits bind ATP but do not participate directly in catalysis. The γ subunit includes a long helical coil that extends into the center of the $\alpha\beta\alpha\beta\alpha\beta$ hexamer. It breaks the symmetry of the hexamer: each β subunit interacts with a different face of the γ subunit and thus has a different conformation. One β subunit is in the loose conformation, which binds ADP and P_i. A second β subunit is in the tight conformation, which binds ATP with great avidity, so much so that it will convert bound ADP and P_i into ATP. Both the tight and loose conformations are sufficiently constrained that they cannot release their bindings. The final subunit is in the open conformation, which can release the bound ATP in response to the rotation of the γ subunit. Importantly, it is the proton flow that drives the rotation of the γ subunit. Therefore, the role of the proton gradient is not to form ATP but to release it from the synthase.

1.6 GLUCOSE STORAGE AND GENERATION

1.6.1 GLYCOGENESIS

Glycogenesis is the process of glycogen synthesis, in which glucose molecules are added to chains of glycogen for storage. This process is activated during rest periods following the Cori cycle, in the liver, and also activated by insulin in response to high glucose levels, for example after a carbohydrate-containing meal.

For we humans, a continuous supply of exogenous glucose is impractical, because we ingest food only discretely — typically three meals a day. Therefore, a good strategy is to store glucose in the form of glycogen during a meal. During the period between two meals, glycogen in the liver can be converted back into glucose to nourish the whole body. Besides glycogen, triglyceride (fat) is also an important store of energy. However, fat must be burned with oxygen, which might become very limited under conditions such as during strenuous exercise. In contrast, glycogen can produce ATP by glycolysis, without the participation of oxygen (anaerobic respiration). Moreover, glucose is virtually the only fuel for the brain, because fatty acids cannot pass through the blood–brain barrier. Therefore, glycogen appears to be more precious than triglyceride.

Glycogenesis depends on the enzyme glycogen synthase, which takes short polymers of glucose and converts them into long polymers. It catalyzes the addition of uridine diphosphate glucose (UDP-glucose) to the nonreducing ends of glycogen molecules. UDP-glucose is formed by the reaction of uridine triphosphate (UTP) and glucose 1-phosphate. As is the case for glycogen breakdown, the glycogen molecule must be remodeled for continued synthesis. The reactions in glycogen synthesis consist of the following steps:

$$\text{Glucose 6-phosphate} \rightarrow \text{Glucose 1-phosphate}, \tag{1.28}$$

$$\text{Glucose 1-phosphate} + \text{UTP} \rightarrow \text{UDP-glucose} + \text{PP}_\text{i}, \tag{1.29}$$

$$\text{PP}_\text{i} + \text{H}_2\text{O} \rightarrow 2\text{P}_\text{i}, \tag{1.30}$$

$$\text{UDP-glucose} + \text{Glycogen}_n \rightarrow \text{Glycogen}_{n+1} + \text{UDP}, \tag{1.31}$$

$$\text{UDP} + \text{ATP} \rightarrow \text{UTP} + \text{ADP}. \tag{1.32}$$

The net reaction is

$$\text{Glucose 6-phosphate} + \text{Glycogen}_n + \text{ATP} + H_2O \rightarrow \text{Glycogen}_{n+1} + \text{ADP} + 2P_i. \quad (1.33)$$

When the plasma insulin concentration is low (e.g., during fasting), glycogen synthase is phosphorylated by glycogen synthase kinase (GSK) and is thus inactive. After a meal, the rising plasma glucose stimulates the pancreatic beta cells to secrete more insulin. The rising insulin activates AKT (also known as protein kinase B) through a molecular circuit known as the PI3K-AKT-TOR pathway (see Chapter 7). The activated AKT inhibits GSK and thus indirectly activates glycogen synthase. Insulin is thus a potent anabolic hormone promoting the synthesis of glycogen.

1.6.2 Glycogenolysis

Glycogenolysis is the breakdown of glycogen to glucose. It consists of the following three steps. First, the enzyme glycogen phosphorylase performs an action called phosphorolysis, namely, the cleavage of a bond of glycogen by adding orthophosphate (P_i) to yield glucose 1-phosphate:

$$\text{Glycogen}_n + P_i \leftrightharpoons \text{Glucose 1-phosphate} + \text{Glycogen}_{n-1}, \quad (1.34)$$

where glycogen$_n$ represents a glycogen with n residues. Glucose 1-phosphate is then released from glycogen. This reaction is readily reversible in vitro. In vivo, however, the reaction is primarily in the direction of glycogen breakdown, because the ratio between [P_i] and [glucose 1-phosphate] is usually greater than 100, which greatly favors phosphorolysis. Compared with a hydrolytic cleavage, the phosphorolytic cleavage is energetically advantageous, because the product is already phosphorylated. This would save one ATP molecule in the following glycolysis. In addition, glucose 1-phosphate is trapped within the cell, because there does not exist the corresponding transporters to wastefully send glucose 1-phosphate out of the cell. Second, the remaining glycogen is remodeled by two enzymes, transferase and α-1,6-glucosidase, so that it can be further degraded by the phosphorylase. Finally, glucose 1-phosphate is converted into glucose 6-phosphate for further metabolism. This reaction, essentially a shift of a phosphoryl group, is catalyzed by phosphoglucomutase. To effect this shift, the enzyme exchanges a phosphoryl group with the substrate. That is, a phosphoryl group is transferred from the enzyme to the substrate, and a different phosphoryl group is transferred back to restore the enzyme to its initial state.

Glucose 6-phosphate can be metabolized to produce ATP. It can also be processed by the pentose phosphate pathway to yield NADPH and ribose derivatives. These activities are what muscles do to process glucose 6-phosphate. Things are a little different for the liver, which has another way (which is more important) to process glucose 6-phosphate (see the next paragraph). Skeletal muscle is a major depot of glycogen. The glycogen concentration in the liver (10%) is higher than in muscles (2%), but more glycogen is stored in skeletal muscles overall because of muscles' much greater mass. Because myocytes lack the enzyme glucose 6-phosphatase, glucose 6-phosphate in the myocytes cannot be converted into glucose and then diffuse out. Glucose 6-phosphate itself, like glucose 1-phosphate, cannot escape from the cell, because the specific transporter does not exist. Therefore, glycogen in the muscle can only be used by the muscle itself.

In the liver, glucose 6-phosphate can be converted into free glucose for release into the bloodstream, which allows the liver to provide endogenous glucose to nourish the whole body. The liver contains glucose 6-phosphatase, a hydrolytic enzyme that is absent in the muscles. This enzyme cleaves the phosphoryl group to form free glucose and orthophosphate, which enables glucose to leave the liver through GLUT2.

Phosphorylase exists in two interconvertible forms: phosphorylase *a*, which is usually active, and phosphorylase *b*, which is usually inactive. Due to its role as the glucose source for other tissues, the liver needs to break down glycogen to produce glucose most of the time. Therefore, phosphorylase *a* is more important than phosphorylase *b* in the liver. After a meal, the liver itself also absorbs glucose, and the rising glucose level in the liver inhibits phosphorylase *a* and prevents the breakdown of glycogen. In contrast, the inactive phosphorylase *b* is dominant in the muscle. This is not surprising because muscles are in the resting state most of the time. During exercise, the AMP level elevates, which leads to the activation of phosphorylase *b*. Exercise also induces the secretion of several hormones, which promote the transition of phosphorylase *b* to phosphorylase *a*. These actions ensure a robust degradation of glycogen in the muscles during exercise. The hormones triggering glycogen breakdown include glucagon and epinephrine. Epinephrine markedly stimulates glycogen breakdown in muscle. The liver is more responsive to glucagon, secreted by pancreatic α cells when the plasma glucose level is low.

1.6.3 Gluconeogenesis

Gluconeogenesis is a metabolic pathway that results in the generation of glucose from noncarbohydrate carbon substrates such as pyruvate, lactate, glycerol, glucogenic amino acids, and odd-chain fatty acids. Like glycogenolysis, it is one of the main mechanisms humans and many other animals use to keep blood glucose levels from dropping too low. Gluconeogenesis is a ubiquitous process, present in plants, animals, fungi, bacteria, and other microorganisms. In vertebrates, gluconeogenesis takes place mainly in the liver and, to a lesser extent, in the cortex of kidneys. In ruminants, this tends to be a continuous process. In many other animals, the process occurs during periods of fasting, starvation, low-carbohydrate diets, or intense exercise.

Gluconeogenesis is necessary because the glucose generated by glycogenolysis is often limited. Because only the liver has the required enzyme glucose 6-phosphatase, only the liver can use the stored glycogen to supply glucose to the blood. Glycogen stored in the muscles, although in large amounts, can be used only by muscles themselves. This limited glycogen storage in the liver (only about 100 grams) is certainly insufficient to provide adequate glucose for a prolonged fasting or starvation. In this event, the liver can generate glucose from noncarbohydrate molecules, a process termed gluconeogenesis, which is promoted by the hormone glucagon as well. Because the liver constitutes only about 3% of body weight, the carbon reservoir in the liver is very limited and is insufficient for gluconeogenesis. Fortunately, muscles can act as a bulk carbon source during starvation, because the low insulin level allows for a rapid degradation of muscle proteins. Amino acids released from muscles can then be used by the liver as the substrate for gluconeogenesis. The conversion of amino acids into pyruvic acids and glucose depends on several enzymes, whose production is stimulated by glucagon and cortisol (an adrenal hormone). During prolonged fasting and exercise, gluconeogenesis in the liver using amino acids from muscles may be the only source of blood glucose.

1.6.4 Fat Synthesis From Glucose

Glycogen represents only a small portion of the stored glucose. In a 70 kg man, glycogen accounts for less than 2000 kcal of stored energy: about 350 g is stored in skeletal muscles to be used only by the muscle themselves; about 90 g is stored in the liver and can be converted to glucose and used by other organs. In contrast, 80% to 85% of the body's energy is stored as fat, which amounts to about 140,000 kcal. Note that 1 g of fat is equivalent to 9 kcal of energy; 1 g of carbohydrates or protein is equivalent to 4 kcal of energy.

In order to be converted into fat, glucose must undergo glycolysis so that pyruvic acid can be converted into acetyl CoA. If acetyl CoA enters the citric acid cycle, then the final outcome is ATP, instead of fat. Alternatively, the acetic acid subunits of these acetyl CoA molecules can be used to produce a variety of lipids, including cholesterol, ketone bodies, and fatty acids. The formation of fatty acids goes as follows. A number of acetic acid subunits (each with two carbons) are joined together to form a chain of fatty acids. When three of these fatty acid chains condense with one glycerol (derived from phosphoglyceraldehyde), a triglyceride (also called triacylglycerol) molecule is produced. The formation of fat, or lipogenesis, occurs primarily in adipose tissue and in the liver when the concentration of plasma glucose is elevated following a meal.

2 Glucose Homeostasis

2.1 HOMEOSTASIS

The Frenchman Claude Bernard (1813–1878) noticed a remarkable constancy in the internal environment (*milieu intérieur*) of a living organism, despite changing conditions in the external environment. He wrote "The constancy of the internal environment is the condition for a free and independent life" (*La fixité du milieu intérieur est la condition d'une vie libre et indépendante*). A living body, as he explained, is relatively independent of the surrounding environment. This independence derives from the fact that the tissues are protected by a veritable internal environment which is constituted by the fluids circulating in the body.

In 1932, the American physiologist Walter Cannon (1871–1945) published a book entitled *The Wisdom of the Body* [6]. In the book, Cannon coined the term *homeostasis* and proposed several propositions to describe the general features of homeostasis.

1. Homeostasis is usually manifested by internal constancy such as glucose concentrations, body temperature, and acid-base balance.
2. Any tendency toward change automatically induces factors that resist the change. For example, the increase of plasma glucose concentration would result in thirst and would consequentially result in water uptake, whereby sugar can be diluted in the extracellular fluid.
3. The many mechanisms of physiological regulation act simultaneously or successively to cooperatively maintain the internal constancy. For example, plasma glucose is regulated by insulin, glucagons, and other hormones that control the release of glucose from the liver or the uptake of glucose by tissue cells.

The maintenance of homeostasis can be reflected by certain physiologic indices such as temperature and blood sugar concentration, most of which can be measured simply by a blood sampling. In the fasting state, for example, the plasma glucose concentration is usually maintained within the narrow range 65–105 mg/dl. Because 1 mg/dl is about 0.0555 mM as far as glucose is concerned, the normal range of plasma glucose concentration is about 4–6 mM. The fasting insulin concentration is usually maintained within the narrow range 5–10 μU/ml. The arterial pH value is usually maintained within the narrow range 7.35–7.45. These normal ranges are the set points, namely, the reference values of the corresponding physiologic indices used by the body, any deviation from a set point is detected by the corresponding sensor in the body, and the information is relayed to an integrating center, which is often a group of cells in the brain or an endocrine gland. The integrating center usually receives information from many different sensors, whose integration leads to a response by either increasing or decreasing the activity of a particular effector (e.g., muscles or glands). Most importantly, the effector aims to reduce the difference between the actual physiologic quantity and the set point consistently. For this reason, the regulation is a negative feedback control. Because strict constancy can never be reached by negative feedback, the set point is actually a narrow range and can be quantified as the median value of the range.

Homeostasis is usually maintained by endocrine systems. Regulation by the endocrine system is achieved by the secretion of hormones into the blood in response to specific stimuli. These hormones are then carried by the blood to all the organs in the body. A rise in the plasma glucose concentration, for example, stimulates insulin secretion from structures in the pancreas known as the pancreatic islets, or islets of Langerhans. Hormones are also secreted in response to nerve stimulation and in response to stimulations from other hormones. The secretion of a hormone can be inhibited by its own effects, following the negative feedback mechanism. Insulin is a well-known peptide hormone that plays the pivotal role in glucose homeostasis. Plasma glucose and insulin, together with their related organs, constitute a negative feedback system, which will be described in detail in the following section.

2.2 GLUCOSE-INSULIN FEEDBACK SYSTEM

2.2.1 GLUCOSE

Glucose is arguably the most important fuel for the body. It can be used to generate ATP with or without oxygen. In contrast, fatty acids can only be burned with oxygen. The brain depends absolutely on plasma glucose as its fuel. Unlike most other organs that can store energy, the brain requires a continuous supply of glucose. In the resting state, the brain consumes about 60% of the glucose supply. Most cell types in the body have no problems in obtaining glucose, due to the varieties of glucose transporters expressed in a cell. GLUT1 is a ubiquitous glucose transporter, which is expressed in all the cells in the body so that every cell can obtain glucose, albeit in small quantity. This low capacity is certainly insufficient for the brain. Thus, neurons primarily use GLUT3 to import glucose, which has much greater transportation capacity than GLUT1. The great transportation capacity of GLUT3 is reflected by the large value of k_{cat}, which is the largest among all the GLUTs. Besides, GLUT3 has a very high affinity with glucose (reflected by the small value of K_m, which is the smallest among all the GLUTs). Some cells (e.g., liver cells, pancreatic beta cells) use GLUT2 to transport large volumes of glucose. Striated (skeletal and cardiac) muscles and adipose tissues are special, because the activation of their major glucose transporter, GLUT4, depends on the peptide hormone insulin. Under the condition of low insulin concentrations, GLUT4 is sequestered in intracellular vesicles in muscle and fat cells. When insulin becomes sufficiently high, the vesicles translocate to the plasma membrane and fuse with the plasma membrane, whereby GLUT4 transporters are inserted onto the plasma membrane and become available for transporting glucose.

2.2.2 INSULIN

Insulin is secreted by the pancreas, an organ that has both exocrine and endocrine functions. The exocrine function is primarily for the digestive system. Scattered in a "sea" of pancreatic exocrine tissues are islands of endocrine cells, called the pancreatic islets or islets of Langerhans. The islets primarily consist of three distinct cell types (alpha, beta, delta), each of which secretes a distinct hormone. The alpha cells, which make up about 25% of the pancreatic islets, secrete the hormone glucagon. When the plasma glucose concentration becomes very low, glucagon stimulates the liver to carry out glycogenolysis, namely, the conversion of glycogen into glucose, so that the plasma glucose can rise to an acceptable level. Similarly, glucagon can stimulate the degradation of triglyceride into free fatty acids and even further into ketone bodies, to sustain the body's needs. Ketone bodies can be utilized by the brain during starvation, when the plasma glucose concentration becomes overly low. The delta cells form about 15% of the pancreatic islets. They secrete the hormone somatostatin, whose function is still not well understood. The beta cells, which make up of about 60% of the pancreatic islets, secrete the hormone insulin in response to a rise in plasma glucose concentration. Insulin secretion is largely stimulated by the plasma glucose concentration and to a lesser degree by the concentration of amino acids. The following discussion will focus on insulin-regulated glucose homeostasis, although other nutrients (amino acids and fat) will also be mentioned at appropriate times.

2.2.3 Glucose and Insulin Homeostasis

In the fasting state, the glucose and insulin concentrations in our blood are maintained at their basal levels, which are within a narrow range: 65–105 mg/dl for glucose [10] and 5–10 μU/ml for insulin [26]. The maintenance of the basal levels depends on the liver, which can secrete glucose to the bloodstream to replenish glucose utilized by key organs in the body. When the glucose level drops too low, pancreatic alpha cells secrete more glucagon, which stimulates the liver to carry out glycogenolysis and gluconeogenesis, both of which are described in detail in Section 1.6.

Following a meal, glucose and insulin concentrations increase for a while and then reduce to the basal levels. The rise of glucose and insulin concentrations is apparently caused by the ingestion of the meal. Starch is the most common carbohydrate in the human diet and is contained in large amounts in such staple foods as wheat, rice, potatoes, corn, and cassava. Because starch is a carbohydrate consisting of a large number of glucose units joined by glycosidic bonds, ingestion of a meal naturally leads to a rise of the plasma glucose concentration. The rise of glucose is sensed by the pancreas, which secretes more insulin. As a consequence, the plasma insulin concentration also rises. Due to glucose utilization by tissue cells, the glucose concentration soon levels off, which is followed by leveling off of the insulin concentration. In the healthy state, the peak glucose concentration upon ingestion is usually around 140 mg/dl (see Figure 3.2(f)) and the peak insulin concentration is usually around 50 μU/ml (see Figure 3.2(e)).

Cells can be roughly divided into two categories according to their dependence on insulin for glucose utilization. The first category includes neurons, whose glucose utilization is insulin independent. They uptake glucose continuously at a relatively constant rate, which is a factor contributing to the leveling off and reduction of the plasma glucose and insulin concentrations. The second category includes myocytes and adipocytes, whose glucose utilization depends on insulin. As the insulin concentration increases in the blood, these cells speed up their glucose uptake. The massive glucose utilization causes a decrease of the plasma glucose concentration. And the insulin level falls accordingly because the pancreatic beta cells become less stimulated. Finally, both glucose and insulin return to their basal levels. Note that we used the words "speed up" to express the common view that insulin-mediated glucose uptake increases gradually as the insulin concentration increases. It will turn out that this assumption, although intuitively natural, may be inaccurate.

Because insulin increase is caused by glucose increase, but the effect of insulin increase is to cause glucose decrease and consequent insulin decrease, the system is essentially a negative feedback system. It is this negative feedback that tightly regulates whole-body glucose-insulin homeostasis, namely, the maintenance of the relative constancy of both concentrations even under the frequent perturbations from meal ingestion.

While some of the absorbed glucose is metabolized immediately to generate ATP, the majority of it is converted into glycogen for storage in the liver and muscles. Being massive, skeletal muscles are responsible for most of the insulin-stimulated glucose storage. For a nonobese man weighing 70 kg, the glycogen storage is about 0.5 kg: 0.1 kg in the liver and 0.4 kg in skeletal muscles. Once the glycogen stores have been filled, the continued ingestion of excess calories results in the production of fat.

Our diets also contain other nutrients such as fat and amino acids. These nutrients are also absorbed into the cells through their respective transporters. The absorbed glucose and fatty acids are either metabolized directly or stored as glycogen (in the liver and muscles) and triglycerides (in adipose tissues). The absorbed amino acids are used to synthesize cellular protein. For a nonobese man weighing 70 kg, the fat storage is about 10 kg (82,500 kcal), which is sufficient for about 40 days of the body's energy expenditure.

2.2.4 Aberrant Homeostasis

Deregulation of the glucose-insulin feedback system may cause prolonged deviation of plasma glucose and insulin concentrations from their normal ranges, which marks serious health problems and increases the risk of many diseases. The essence of these diseases will be discussed in detail in Part III of the book. In the following, we briefly describe the hallmarks of some of these diseases.

Hyperglycemia, or high blood sugar, is a condition in which an excessive amount of glucose circulates in the blood plasma. It is usually quantified as a glucose level higher than 200 mg/dl. But symptoms may not start to become noticeable until even higher values such as 250–300 mg/dl (i.e., 15–20 mM) are reached. Hyperglycemia is a hallmark of diabetes. A glucose level above 126 mg/dl is used as a criterion for the diagnosis of diabetes. Such a chronic high level of glucose can damage tissues and organs. A well-known example is damage to the eyes and related vision problems. A high plasma glucose concentration harms four parts of the eye — the retina, the vitreous fluid, the lens, and the optic nerve. The most common eye problem is diabetic retinopathy. As this condition develops, the blood vessels of the eye begin to weaken. Blood may leak and block light from hitting the retina. Damaged blood vessels cause scar tissue to grow; the tissue may eventually pull the retina back from the eye and block light.

Hypoglycemia, or low blood sugar, is a condition in which there is an abnormally diminished content of glucose in the blood. It can produce a variety of symptoms and effects but the principal problems arise from an inadequate supply of glucose to the brain, resulting in impairment of function (neuroglycopenia). Effects of hypoglycemia can range from mild dysphoria to more serious issues such as seizures, unconsciousness, and permanent brain damage or death. The most common forms of hypoglycemia occur as a complication of the treatment of diabetes mellitus with insulin or oral medications. Hypoglycemia is less common in nondiabetic persons, but can occur at any age. Among the causes of hypoglycemia are prolonged starvation, alterations of metabolism associated with infection, organ failure, and hyperinsulinemia.

Hyperinsulinemia is a condition in which there are higher levels of insulin circulating in the blood than expected relative to the level of glucose. While hyperinsulinemia is often seen in people with early-stage type 2 diabetes mellitus, it is not the cause of the condition and is only one symptom of the disease. Hyperinsulinemia is associated with conditions such as hypertension, obesity, dyslipidemia, and glucose intolerance, which are collectively known as metabolic syndrome. This close association between hyperinsulinemia and metabolic syndrome suggests related or common mechanisms of pathogenicity. Hyperinsulinemia also increases the risk of certain types of cancer, because insulin is a hormone favoring growth.

To better understand aberrant phenomena in glucose and insulin homeostasis, it is worthwhile to investigate the fine details of the glucose-insulin feedback system by mathematical modeling. This will be the task of the next section.

2.3 MATHEMATICAL MODELING

In this section, a mathematical model is used to describe the glucose-insulin feedback system. In the model, the symbol t has the usual meaning of time; $G(t)$ and $I(t)$ represent the temporal changes of plasma glucose and insulin concentrations. The model consists of the following ordinary differential equations:

$$\frac{dG(t)}{dt} = s + m(t) - \lambda G(t) - u(I(t))\,G(t), \tag{2.1}$$

$$\frac{dI(t)}{dt} = f(G(t)) - kI(t), \tag{2.2}$$

where

t = time.
$G(t)$ = glucose concentration in the blood.
$I(t)$ = insulin concentration in the blood.
s = glucose source rate supplied by the liver, less the approximately constant rate of glucose utilization by the brain.
$m(t)$ = rate of glucose supplied by a meal; one has $m(t) \equiv 0$ during fasting. See below for a more detailed description of $m(t)$.
λ = rate of insulin-independent glucose utilization per unit glucose concentration. For example, glucose transmitted through GLUT1 contributes to λ.

$u(I)$ = rate of insulin-dependent glucose utilization (primarily by skeletal muscle cells and adipocytes) per unit glucose concentration, which is the main factor controlling plasma glucose clearance.
k = degradation rate of insulin.
$f(G)$ = rate of insulin production by pancreatic beta cells in response to glucose stimulation; it is a monotonically increasing function of the glucose concentration. See below for a more detailed description of $f(G)$.

The aim of mathematical modeling is to study insulin responsiveness, namely, the determination of $u(I)$. It is natural to think that $u(I)$ is a continuous and monotonically increasing function (the more insulin, the faster glucose uptake), which has been assumed by others (e.g., [30]). This book uses a completely different approach. We do not assume the function form of $u(I)$ but treat it as unknown. The aim is to determine the specific $u(I)$ that renders *optimal* glucose homeostasis, with some fairly generic optimality criteria. The practicality and biological relevance of the obtained optimal control will be discussed in Part II of this book, where the actual control rendered by molecular circuits will be determined and compared with the optimal control.

2.3.1 The Function $m(t)$

Let $t = 0$ be the time when meal ingestion begins and let $m(t)$ be the rate of glucose supplied by the meal (Figure 2.1). Upon meal ingestion, $m(t)$ rises initially, plateaus, and then declines gradually. Both the rising and the declining phases of $m(t)$ can be modeled by sigmoidal functions. Because $m(0) = 0$, for the rising phase we use a Hill function. Because $m(t) \to 0$ as $t \to \infty$, for the declining phase we use a sigmoidal function whose kernel is an exponential function exp(·). The function $m(t)$ is taken to have the following form:

$$m(t) = \begin{cases} M(t/\tau_1)^b / \left(1 + (t/\tau_1)^b\right) & \text{for } 0 \le t < \tau_2 \\ M \cdot \exp(c(\tau_3 - t)) / (1 + \exp(c(\tau_3 - t))) & \text{for } \tau_2 \le t < \infty \end{cases}$$

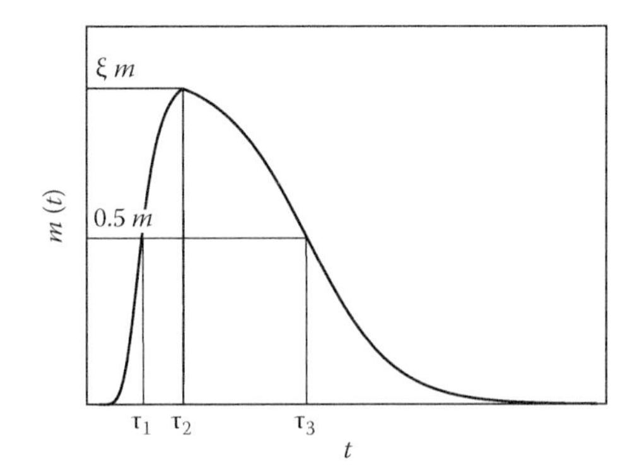

FIGURE 2.1 The function $m(t)$, the rate of glucose supplied by a meal, is modeled by a Hill function and an exponential function. It rises initially, plateaus, and then declines gradually to zero.

where

$$b = \frac{\ln(\xi/(1-\xi))}{\ln(\tau_2/\tau_1)}$$

and

$$c = \frac{\ln(\xi/(1-\xi))}{\tau_3 - \tau_2}.$$

The function $m(t)$ is completely determined by τ_1, τ_2, τ_3, ξ, and M. The parameter $\xi < 1$ provides a continuous joining of the two phases of $m(t)$ through modulating b and c. The closer to 1, the smoother the joint. The parameter M determines the magnitude of $m(t)$. The parameter τ_2 is the time when $m(t)$ peaks. The parameter τ_1 (τ_3) is the half time of the rising (declining) phase of $m(t)$.

2.3.2 The Function $f(G)$

The function $f(G)$ represents the rate of pancreatic insulin secretion in response to glucose, which is a monotonically increasing, sigmoidal function of glucose concentration. It is therefore modeled by a Hill function (Figure 2.2):

$$f(G) = f_{\max}\frac{(G/G_h)^n}{1+(G/G_h)^n},$$

where $f_{\max}$ is the maximum rate of insulin secretion, G_h corresponds to the glucose level at which the insulin secretion rate is $f_{\max}/2$, and n is the Hill coefficient.

2.3.3 Fasting State

During fasting, there is no exogenous glucose ($m(t) \equiv 0$) and no insulin action ($u(I) \equiv 0$), and the mathematical model can be simplified as

$$\frac{dG(t)}{dt} = s - \lambda G(t), \tag{2.3}$$

$$\frac{dI(t)}{dt} = f(G(t)) - kI(t). \tag{2.4}$$

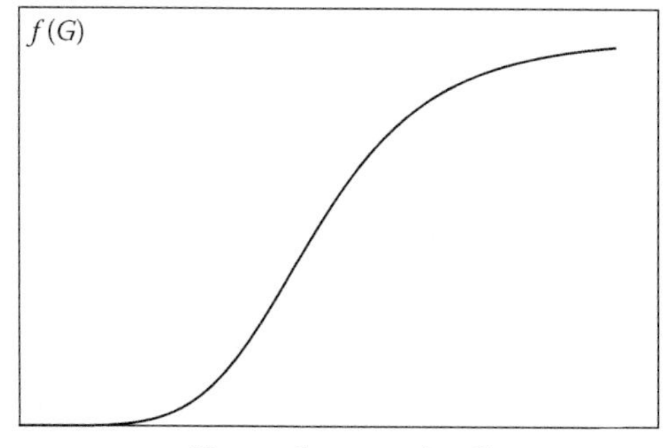

FIGURE 2.2 The function $f(G)$, the rate of insulin production in response to glucose stimulation, is modeled by a Hill function.

The system maintains at a steady state (G_0, I_0) during fasting, where

$$G_0 = \frac{s}{\lambda} \tag{2.5}$$

and

$$I_0 = \frac{f(s/\lambda)}{k} \tag{2.6}$$

are the plasma glucose and insulin concentrations, respectively. Note that Equations (2.5) and (2.6) are obtained by applying $dG/dt = 0$ and $dI/dt = 0$. Because the normal ranges of fasting glucose and insulin concentrations are 65–105 mg/dl and 5–10 μU/ml, respectively, we set $G_0 = 85$ mg/dl and $I_0 = 7.5$ μU/ml. This implies the following relationship among the system parameters: $s = 85\lambda$ and $k = f(85)/7.5$.

2.3.4 Absorptive State

The food intake $m(t)$ induces a rapid increase of the plasma the glucose concentration $G(t)$. The rising glucose stimulates the pancreas to secrete more insulin. This is reflected by the monotonically increasing function $f(G)$ in the model: the larger G is, the larger the value of $f(G)$. In response to the rising insulin level, skeletal muscles and adipose tissues actively uptake glucose through a series of actions at the molecular level. These molecular events will be described in detail in Chapter 7. In brief, insulin activates the protein kinase AKT, which promotes the translocation of GLUT4 from an intracellular store to the plasma membrane; GLUT4 then carries glucose into the cell, whereby the plasma glucose level is lowered. The rate of insulin-mediated glucose uptake is recapitulated by the mass action term $-u(I)G$ in the mathematical model. Following the entry of a large portion of glucose into the tissue cells, the insulin level falls because pancreatic beta cells become less stimulated. Finally, both glucose and insulin return to the steady states, namely, their basal levels G_0 and I_0. Let T denote the time when homeostasis is first restored. The smaller T is, the more efficient the insulin action.

The above narrative describes the glucose and insulin dynamics in response to the ingestion of one meal. For a healthy person, the dynamical response repeats three times a day following three meals. The green dots connected by solid lines in Figure 3.2(f) represent clinical data of 24-h profiles of plasma glucose and insulin concentrations averaged from 14 normal subjects. Figure 3.2(e) presents the corresponding insulin concentrations. The three spikes in both figures at 09.00, 13.00, and 18.00 are caused by the three meals. One sees that the glucose and insulin dynamics are largely in parallel. It demonstrates that the change of insulin concentration tracks closely the change of glucose concentration, which is attributable to the adaptive response of the pancreas.

In Chapter 3, numerical simulations of the mathematical model will be performed. Figures 3.2(c) and 3.2(b) illustrate a simulation run, with the clinical data in Figures 3.2(f) and 3.2(e) placed alongside as a comparison. One sees that the simulation mimics the clinical data well. In particular, both simulations and the actual data indicate that insulin dynamics track well with glucose dynamics, demonstrating the great parallelism between glucose and insulin.

3 Optimal Glucose Homeostasis

3.1 OPTIMIZATION AND HOMEOSTASIS

Optimization is a fundamental principle in biology. Aristotle once stated: "If one way be better than another, that you may be sure is nature's way" [31]. Darwin's theory of natural selection provided an obvious mechanism for optimization in biology: more optimized individuals are more likely to survive. After many rounds of natural selection, a species is optimally adapted to the environment in which it evolved. Optimality is thus absolutely necessary for the survival of a species — it is not something luxurious for making one's life sweeter. The optimality principle may allow scientists to predict how organisms should be organized, besides merely to describe patterns or mechanisms.

Homeostasis, as a central concept in biology, is closely related to optimization. An advantage of homeostatic regulation is that it allows an organism to function effectively in a broad range of environmental conditions. For example, an ectotherm tends to become sluggish at low temperatures, whereas an endotherm in the neighborhood may be fully active. Incidentally, an ectotherm is an organism in which internal physiological sources of heat are of relatively small or quite negligible importance in controlling body temperature; and an endotherm is an organism that maintains its body at a metabolically favorable temperature, largely by use of heat set free by its internal bodily functions instead of relying almost purely on ambient heat. Homeostasis is usually maintained by endocrine systems. Regulation by the endocrine system is achieved by the secretion of hormones into the blood in response to specific stimuli. These hormones are then carried by the blood to all the organs in the body. The secretion of a hormone can be inhibited by its own effects, in a negative feedback manner. This closed-loop control system is called negative feedback inhibition. The negative feedback control must accommodate competing or even conflicting internal requirements and possibly wide environmental variations. Those that achieve optimal compromise are advantageous to survive natural selection. Optimal homeostatic regulation appears to be an overarching principle acquired through evolution.

Glucose homeostasis is one of the most important homeostatic processes in biology. It was an example frequently used by W. B. Cannon to elucidate homeostasis. Unicellular organisms depend relatively less on homeostatic regulation — a cell just takes in nutrients consecutively until the cell mass doubles; cell division then ensues. The evolution of multicellularity necessitates a well-balanced nutrient distribution, because different cell types may have different energy requirements. This is especially true for advanced organisms such as animals, because an ordered glucose distribution is necessary to ensure more important organs get more glucose or get glucose earlier. Neurons, relying absolutely on glucose as their energy source, need to absorb glucose in an autonomous manner. Skeletal muscles, being less important than the brain, should have their glucose uptake controlled. It is desired that muscle cells remain firmly closed when the plasma glucose concentration is low (e.g., during fasting) so that the limited glucose source can be spared for the brain. Because proteins are more suitable to act as signaling molecules than are carbohydrates, nature developed insulin, a peptide hormone, to closely track glucose change and act as the control signal. It is desired that muscle cells possess a certain "insulin resistance" so that they absorb glucose only after the plasma insulin level has become sufficiently high. In this sense, insulin acts more as a factor *delaying* glucose entry than as a factor blindly promoting glucose entry.

On the other hand, it is desired that skeletal muscles are highly responsive to insulin so that a high glucose concentration in the blood, especially after a meal, can be rapidly removed to avoid hyperglycemia and the consequent toxicity. If plasma glucose is maintained at a concentration above normal for too long, tissue damage will occur. A well-known example is damage to the eyes and the related vision problems.

Therefore, the two conflicting requirements, insulin resistance and rapid insulin responsiveness, have to be fulfilled. The insulin responsive element in the cell, namely, the PI3K-AKT-TOR pathway, is certainly sitting on the horns of a dilemma. The dilemma is especially acute for humans, whose brain size in relation to body weight is the largest among all the species. As a general approximation, the brain size is proportional to the body weight. In the past two million years, the hominid lineage leading to modern humans evolved significantly larger and more sophisticated brains than other primates. The human brain is about 3.5 times larger than that of chimpanzees (*Pan troglodytes*), which are similar in lean body weight to us. The modern human brain is also approximately three times larger than in the now extinct but smaller very early human ancestors, the Australopithecines. The modern human brain is now approximately 6% smaller than it was in the early modern homo sapiens of 25,000–100,000 years ago but we are still equally "encephalized" because our body weight is also a little less than theirs. Such a large brain certainly demands a great deal of encephalic glucose consumption. In adult humans, the brain weighs approximately 2.5% of the body weight but it uses approximately 25% of the body's daily energy requirement.

Unfortunately, food scarcity is a common theme during evolution. The big energy demand of the brain can only be met through fierce competition, and a long period of waiting is often needed. The intense competition for energy often leads to an evolutionary arms race between prey and predator. Although the predator became more competent, the prey evolved new strategies to evade attack. Life never becomes easy for almost every species. Exceptions do exist, such as people like us who are living in a time of unprecedented food abundance. However, that exception only came very recently, due to scientific and technical advances in agriculture, which have yielded an era in which harvests are now outpacing population growth. Our ancestors (hunter-gatherers) were not as fortunate as us. They had to deal with sporadic food availability — a sudden surfeit of food followed by a prolonged starvation. During prolonged starvation (several days without any food), stringent insulin resistance was necessary to prevent muscles from utilizing glucose, so that glucose could be spared for the brain. Upon a sudden surfeit of food ingestion, efficient insulin action was desired to remove glucose from the blood as soon as possible. The alternating periods of food surfeit and food scarcity imposed another constraint on the optimization of glucose homeostasis, in addition to the large energy requirement of the brain.

There does exist an optimal solution that can reconcile the opposing requirements in glucose homeostasis. By using mathematical analysis, particularly an optimal control theory, I found that optimal homeostasis necessitates a bistable control mediated by insulin. The bistability is characterized by all-or-none and hysteresis.

3.2 FORMULATION OF THE OPTIMIZATION PROBLEM

We aim to determine the optimal insulin action $u(I)$ in order to best reconcile the two opposing requirements in insulin signaling: high insulin efficiency to confer rapid plasma glucose clearance and an insulin sparing state to guarantee the brain's safety during fasting. The insulin action $u(I)$ is present in the model Equations (2.1) and (2.2), where u is the rate of plasma glucose reduction per unit glucose concentration and $u(I)$ signifies its dependence on insulin. From the perspective of whole-body glucose homeostasis, it is a negative feedback control that reduces plasma glucose toward its basal level. Because it is the activated AKT (pAKT) that effects glucose uptake, the physiologic quantity u corresponds to [pAKT], the intracellular concentration of the activated AKT. Let $u_{\max}$ be the upper bound of u, which then corresponds to $[\text{AKT}]_{\text{tot}}$, the intracellular concentration of the total AKT. Indeed, the amount of pAKT can approach but not exceed the amount of total AKT. The protein kinase AKT is very stable and $[\text{AKT}]_{\text{tot}}$ is fairly constant for most cell types. Thus, $u_{\max}$ is a constant in this book.

The objective function for optimization is designed as follows. First, the efficiency of insulin action can be represented by the value of T, the time when homeostasis is restored upon a meal perturbation beginning at time $t = 0$. A smaller value of T means that the insulin action is more efficient. Therefore, the optimization should be in terms of the minimization of T. Second, the optimization is constrained by a fixed average:

$$\overline{u} = \frac{1}{T}\int_0^T u(t)dt.$$

By setting $\overline{u}$ significantly smaller than $u_{\max}$ (the upper bound of $u(I)$), the overall action of insulin is limited, which is necessary for brain safety. One can also use $\eta = \overline{u}/u_{\max}$, termed the utilization rate, to set the constraint. The utilization rate η of normal cells should be small. A much larger utilization rate may be adopted by cancer cells, which must consume a large amount of glucose for them to become rapidly dividing. In summary, the aim of the optimization is to find the control $u(t)$ that restores homeostasis in the shortest time, whose average is fixed at $\eta u_{\max}$ for some small η. That is,

$$\begin{aligned} &\min_{u(t)} T \quad \text{subject to} \\ &\frac{1}{T}\int_0^T u(t)dt = \eta u_{\max}. \end{aligned} \tag{3.1}$$

3.3 DETERMINATION OF OPTIMAL SOLUTIONS

Modern control theories have provided standard and effective approaches to determine an optimal control, which can be summarized as the following three steps in the context of obtaining $u(I)$:

- Determine a time function $u(t)$ that is optimal, without considering how it is realized by the control signal I.
- Numerically solve Equations (2.1) and (2.2) with the optimal $u(t)$ to obtain the corresponding $G(t)$ and $I(t)$.
- Construct the optimal control $u(I)$ from the optimal $u(t)$ and the corresponding $I(t)$. For every time point t, one finds the corresponding $u(t)$ and $I(t)$; a point $(I(t), u(t))$ is then traced out on the I versus u plane.

In brief, one first determines the optimal open-loop control $u(t)$ and then determines the closed-loop control $u(I)$.

Note that the optimal control thus obtained is usually only locally optimal. A locally optimal control $u(t)$ is the best one within a neighborhood of the possible optimization variables. In terms of the present problem, a locally optimal control $u(t)$ achieves faster homeostasis than $u(t) + \varepsilon \cdot \delta u(t)$ with any $\delta u(t)$ and $\varepsilon \to 0$.

Once the locally optimal controls are determined, one can further search for the globally optimal control among the local optima. However, it should be stressed that the determination of locally optimal controls is practically more important. First, the exact global optimum bears little meaning for a practical system which is under random perturbations (the system parameters always change). That is, the status of global optimality only applies to a particular set of parameters, and the control solution loses the status once the system parameters change. Second, the information about local optima might have revealed strategies or organizing principles of the system that confer optimality, which might be more important than the global optimum per se.

3.4 LOCAL OPTIMALITY

The glucose-insulin system is subject to certain fluctuations of the parameters s, λ, and k, the shapes of $m(t)$ and $f(G)$, etc. At least, the meal ingestion $m(t)$ differs every time. Therefore, the discussion of optimality should take parameter variations into account. In the following, locally optimal controls will be computed by using symbols, instead of specific values, for the parameters. Such symbolic computations ensure that the obtained local optima are general and are largely independent of the choices of the parameters.

A locally optimal control is usually associated with certain parameter conditions. Let $\mathbf{p}$ abbreviate the system parameters, including s, λ, k, and the parameters used to determine $m(t)$ and $f(G)$. There exists a type of condition termed the singular condition, which is held by a particular set of parameters $\mathbf{p}$ but is not held by any others in the vicinity of $\mathbf{p}$. That is, the singular condition is held by $\mathbf{p}$ but is not held by any $\mathbf{p} + \varepsilon \cdot \delta\mathbf{p}$ (for $\varepsilon \neq 0$ and $\delta\mathbf{p} \neq 0$), no matter how small ε is. If a locally optimal control is associated with a singular condition, then it is not stably optimal, because an infinitesimal perturbation of the parameters would immediately remove the optimality. In more detail, a control $u(t)$ falls into one of the following four categories according to how its optimality depends on a given singular condition **S**.

Category (1): $u(t)$ is locally optimal under any parameter conditions (including optimality on a boundary of the parameter space).
Category (2): $u(t)$ is locally optimal under any parameter conditions except **S**.
Category (3): $u(t)$ is locally optimal only under the singular condition **S**.
Category (4): $u(t)$ is not locally optimal under any parameter conditions.

In this book, a control satisfying category (1) or (2) is considered locally optimal. For category (3), the control is not truly optimal because **S** is singular.

3.4.1 Pontryagin's Maximum Principle

Pontryagin's maximum principle is a celebrated optimal control theory used to find the best possible control for taking a dynamical system from one state to another, especially in the presence of constraints for the state or input controls. It was formulated in 1956 by the Russian mathematician Lev Semenovich Pontryagin and his students [25]. There are two points to be stressed. First, like most optimal control theories, Pontryagin's maximum principle is used to determine locally optimal controls, not the global one. Second, Pontryagin's maximum principle in general only gives *necessary* conditions for optimality. For example, if some feature is derived from Pontryagin's maximum principle, then *any* optimal control must have the feature. However, there may exist controls having the feature but that are not optimal. In Appendix A, Pontryagin's maximum principle is applied to problem (3.1), namely, the determination of insulin action that renders optimal glucose homeostasis. The application is successful and has yielded several impressive results. These results are explained in the following sections. For technical details, readers are referred to Appendix A.

3.4.2 The Optimal Control Is Necessarily All-or-None

By applying Pontryagin's maximum principle, the locally optimal control $u(t)$ is found to be of a bang-bang type — switching abruptly between two extreme values 0 and u_{max}, in a similar way as a residential thermostat in response to a temperature change. That is, the control either turns on completely or turns off completely; it does not take any intermediate values between 0 and u_{max}. Because it is the protein kinase AKT that promotes cellular glucose uptake (equivalently, plasma glucose clearance) in the majority of tissue cells, this optimality imposes an all-or-none requirement at the molecular level: either no AKT activation or full AKT activation.

All-or-none is a recurring phenomenon in biology, which was noticed as early as 1871 by Henry Pickering Bowditch while he was studying the contraction of heart muscle. According to Dr. Bowditch: "An induction shock produces a contraction or fails to do so according to its strength; if it does so at all, it produces the greatest contraction that can be produced by any strength of stimulus in the condition of the muscle at the time" [5]. Bowditch also found that the individual fibers of both skeletal muscle and nerve respond to electrical stimulation according to the all-or-none principle. As the present study shows, skeletal muscles have an all-or-none response not only to an electrical signal (action potentials), but also to a chemical signal (insulin).

More importantly, this all-or-none result does *not* depend on the parameters in the mathematical model (Equations (2.1, 2.2)). As can be found in Appendix A, computations in this part involve only symbols. No values were assigned to the parameters, including the parameters constraining the functions $f(G)$ and $m(t)$. Even the functional forms for $f(G)$ and $m(t)$ were not assumed in the computation. The mathematical model thus represents a general homeostatic system. In Chapter 4, it will further be explained why the mathematical model is general.

Therefore, all-or-none must be a general property. First, homeostasis is a fundamental notion in biology and Equations (2.1, 2.2) represent a general homeostatic system. Second, the all-or-none property is obtained by optimization, and optimality is again fundamental in biology. In conclusion, all-or-none must be a ubiquitous property and its emergence during evolution is inevitable.

3.4.3 The Optimal $u(t)$ Is a Square Pulse

Further analysis revealed that the optimal $u(t)$ is a square wave with only one pulse (Figure 3.1), which has the following expression:

$$u(t) = \begin{cases} 0 & \text{for } 0 \le t < t_{\text{on}} \\ u_{\max} & \text{for } t_{\text{on}} \le t < t_{\text{off}} \\ 0 & \text{for } t_{\text{off}} \le t < T, \end{cases} \tag{3.2}$$

where t_{on} and t_{off} are the time points at which $u(t)$ switches. A control satisfying Equation (3.2) is often abbreviated 0-$u_{\max}$-0 in this book. Being a square pulse, the optimal $u(t)$ has a simple expression for the utilization rate,

$$\eta = \frac{t_{\text{off}} - t_{\text{on}}}{T}.$$

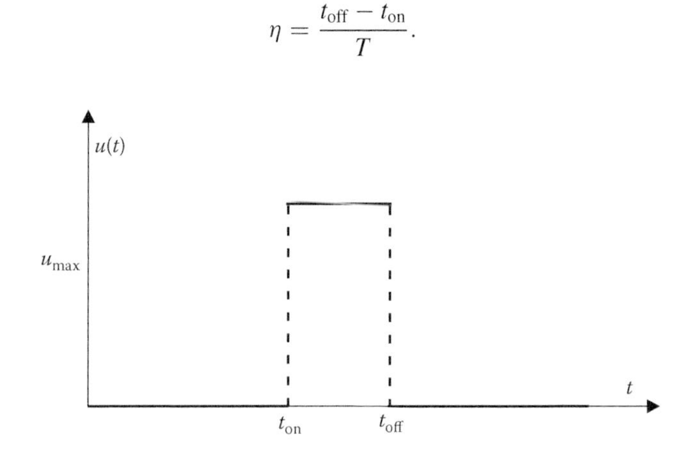

FIGURE 3.1 The optimal $u(t)$ is a square pulse that switches on at a time t_{on} and switches off at a time t_{off}.

The biological meaning of the utilization rate is also clear: it is the duty cycle of glucose uptake.

Like the all-or-none property, the square pulse result does *not* depend on the parameters and functional forms in the mathematical model (Equations (2.1, 2.2)). The obtained local optimality is likely to apply to the homeostasis of all such systems using negative feedback to control concentrations.

We have determined that an optimal control must be 0-u_{max}-0. Conversely, does it hold true that an arbitrary 0-u_{max}-0 control must be optimal? As a matter of fact, a 0-u_{max}-0 control is optimal when a parametric condition (Equation (A.35)) is satisfied. As explained in Appendix A, expression (A.35) should always hold when the model parameters are reasonable. In the following section, some 8.3×10^5 0-u_{max}-0 controls will be tested to determine whether they are optimal controls or not; I found that they all satisfy expression (A.35). Therefore, an arbitrary 0-u_{max}-0 control is almost certainly an optimal one. This implies that local optimality is easy to maintain (the change of t_{on} and t_{off} will not remove optimality). This would allow the system to easily resettle into another optimum when metabolic conditions change.

3.4.4 Determination of the Optimal Control $u(I)$

Because the control of glucose homeostasis is mediated by insulin, the optimal control should finally be expressed in terms of the insulin concentration I. As outlined before, this entails the synthesis of $u(I)$ from $u(t)$ and $I(t)$, where $u(t)$ is the optimal control as a time function (and we already know it is a square pulse), $I(t)$ is the *corresponding* insulin dynamics (which is obtained by integrating Equations (2.1, 2.2) with a square pulse input $u(t)$). The process is analogous to the synthesis of a circle $x^2 + y^2 = 1$ with $x = \cos(t)$ and $y = \sin(t)$.

Figure 3.2 gives an example of the synthesis of the optimal control $u(I)$. The to-be-converted $u(t)$, with $t_{on} = 70.5$ min and $t_{off} = 90.95$ min, is illustrated in Figure 3.2(a). With this $u(t)$, Equations (2.1, 2.2) are integrated to obtain the glucose dynamics $G(t)$ and the insulin dynamics $I(t)$. The values for the model parameters are obtained from the literature and are presented in Table 3.1. Figures 3.2(c) and 3.2(b) show the resultant $G(t)$ and $I(t)$, respectively. One sees the two dynamics are actually in parallel—the change of insulin concentration closely tracks that of glucose concentration. The meal ingestion begins at $t = 0$, which induces a drastic rise of the plasma glucose concentration, followed by an equally sharp increase of plasma insulin (secreted by pancreatic beta cells that are tonically stimulated by glucose). The glucose concentration levels off and then decreases, due to both insulin-mediated glucose utilization and diminishing of the food source. The insulin concentration also decreases, because the pancreas becomes less stimulated. Finally, both concentrations return to their basal levels G_0 and I_0. The results are realistic, as demonstrated by the comparison with clinical data (Figures 3.2(f, e), which are adapted from [24]. The clinical data recorded 24-h glucose and insulin concentrations; the three spikes correspond to the three meals.

As a comparison, the uncontrolled dynamics $\hat{G}(t)$ and $\hat{I}(t)$ are also obtained by setting $u(t) \equiv 0$ during the integration of Equations (2.1, 2.2). The results are shown as the grey curves in Figures 3.2(c, b). One sees that the glucose and insulin

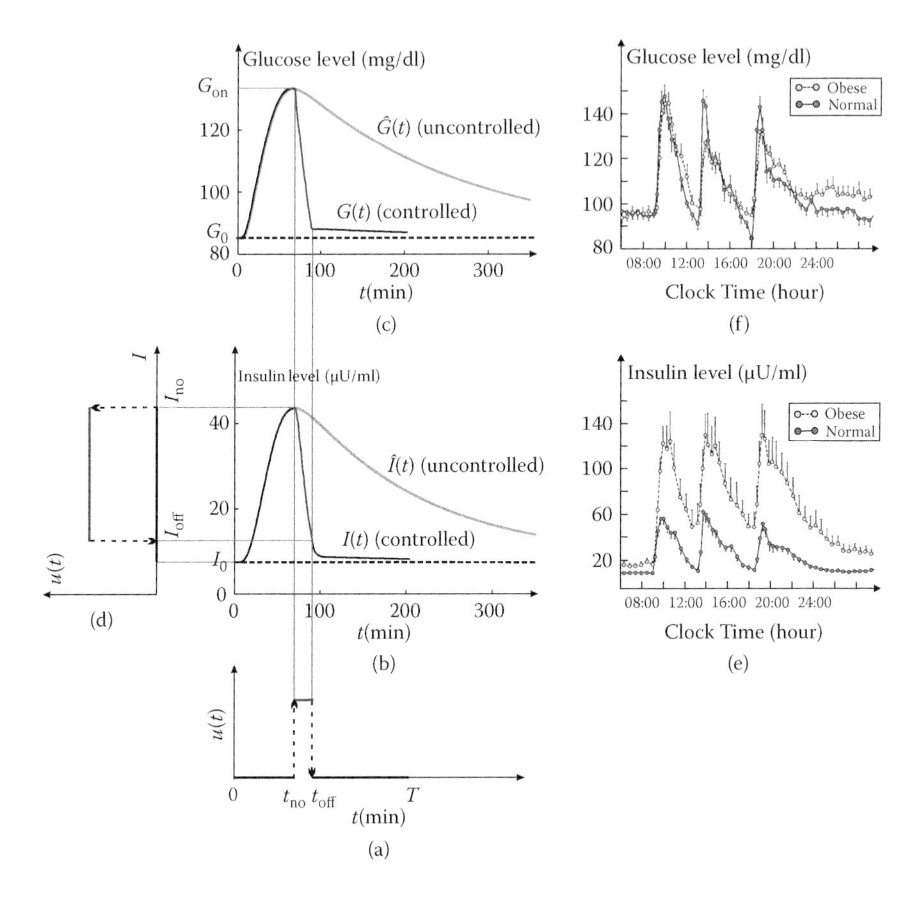

FIGURE 3.2 A computer simulation of the glucose-insulin system. (a) An optimal control as a time function $u(t)$. Note the ON (OFF) phase of the control. (b) The controlled (black) and uncontrolled (grey) insulin dynamics. (c) The controlled and uncontrolled glucose dynamics. (d) The control in the form of $u(I)$. (e) Twenty-four-hour profile of plasma insulin concentration averaged from 14 normal (circles with solid lines) and 15 obese (circles with dashed lines) subjects. The three spikes are caused by the three meals at 09.00, 13.00 and 18.00. (f) Twenty-four-hour profile of plasma glucose concentration. (Parts e and f adapted from [22])

concentrations can still reduce to the basal level G_0 and I_0, due to insulin-independent glucose utilization (the term $-\lambda G(t)$ in the mathematical model). However, the decay rate is much smaller than the controlled dynamics.

3.4.5 The Optimal Control $u(I)$ Is a Bistable (Toggle) Switch

Figure 3.2 has given us a good example of how to determine the optimal control $u(I)$. It can be traced out on the I-u plane as follows. At time t, one obtains the value of $I(t)$ from Figure 3.2(b) and the value of $u(t)$ from Figure 3.2(a). One then draws the point $(I(t), u(t))$ on the I-u plane. By repeating the process for every time point, a curve is traced out that represents $u(I)$ (see Figure 3.2(d)).

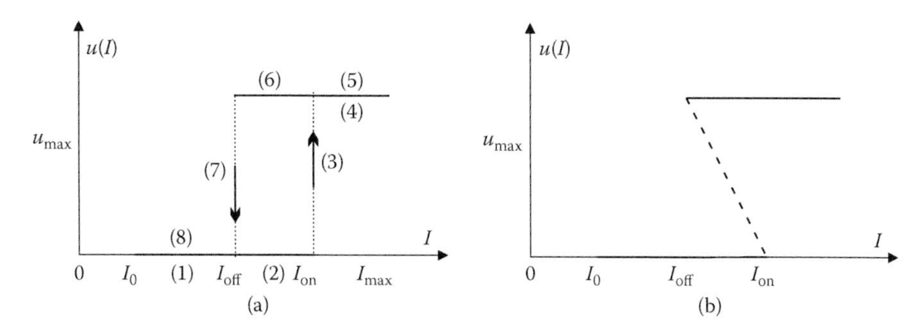

FIGURE 3.3 The optimal control $u(I)$ is a toggle switch. (a) The route of action. (b) A simpler representation.

For a clearer presentation, the I-u plane is rotated 90° clockwise (Figure 3.3(a)). One sees that the obtained optimal control $u(I)$ is a bistable switch that is characterized by all-or-none and hysteresis. Instead of a graded response to the insulin stimulus, the control toggles between the maximum value u_{max} and 0 (all-or-none), but the activation threshold I_{on} significantly differs from the deactivation threshold I_{off} (hysteresis). As I increases, the glucose uptake switches on fully, once I exceeds the threshold $I_{on} = I(t_{on}) = 43.7\ \mu$U/ml. As I decreases, the glucose uptake switches off fully, once I drops below the threshold $I_{off} = I(t_{off}) = 12.23\ \mu$U/ml. Interestingly, hysteresis emerges during the switching: glucose uptake is triggered (terminated) by a larger (smaller) insulin concentration ($I_{on} > I_{off}$). In the following, Figure 3.3(b) will primarily be used to represent a bistable switch. Note that the dashed line in Figure 3.3(b) is just for illustration purposes; it does not represent the change of u as I changes.

Let's assume that the insulin concentration increases from the basal level I_0 to some large value $I_{max} > I_{on}$, and then decreases from I_{max} back to I_0. The control works through the following eight steps, which are also marked in Figure 3.3(a):

1. The control remains at zero as I increases from I_0 to I_{off}.
2. The control remains at zero as I increases from I_{off} to I_{on}.
3. The control switches to u_{max} at $I = I_{on}$.
4. The control remains at u_{max} as I increases from I_{on} to I_{max}.
5. The control remains at u_{max} as I decreases from I_{max} to I_{on}.
6. The control remains at u_{max} as I decreases from I_{on} to I_{off}.
7. The control switches to 0 at $I = I_{off}$.
8. The control remains at 0 as I decreases from I_{off} to I_0.

Although intuitively natural, standard assumptions about $u(I)$ (continuity, monotonicity, and the property of being a function) (e.g., [30]) are challenged by the locally optimal solution obtained here. First, all-or-none implies discontinuity. Second, hysteresis defies monotonicity. Finally and most notably, an optimal control is *not* a function: in the range $I_{off} < I < I_{on}$, one preimage corresponds to two images. Thus, the study on local optimality has already provided important insights into the organizing principles of our body — we "were designed" more nonlinearly than we previously thought.

3.4.6 The Threshold I_{on} and Its Heterogeneity

In Figure 3.2, the control is so powerful ($u_{max} >> \lambda$) that the insulin concentration reduces immediately after the control switches on (i.e., I exceeds the threshold I_{on}). That is, stages (4) and (5) described above are so brief that they are negligible ($I_{max} \approx I_{on}$). It should be stressed that this scenario is not uncommon in biochemistry. Let's assume all the muscle cells have the same I_{on} value (strict homogeneity). Because muscles are so massive (they constitute the main part of the body weight), u_{max} should be larger than λ by several orders of magnitude, because these cells uptake glucose simultaneously and in great quantity. Such a large and simultaneous removal of glucose would certainly lead to an immediate reduction of the plasma glucose level, which is followed by an immediate reduction of the plasma insulin level. That is, I_{max} should be only slightly greater than I_{on}.

Now consider the more realistic case that a certain heterogeneity exists among the muscle cells (Figure 3.4). In this case, the majority of muscle cells (represented by the black switch in Figure 3.4(a)) should have the same I_{on}, at which the insulin concentration reverses ($I_{max} \approx I_{on}$). There are some cells having $I_{on} < I_{max}$ (see the red switch in Figure 3.4(a)); these cells have observable stages (4) and (5) and thus absorb glucose a little longer. There are some cells having $I_{on} > I_{max}$ (see the blue switch in Figure 3.4(a)); these cells do not uptake glucose through GLUT4 because they have not been activated. Indeed, from Figure 3.4(b) one sees clearly that the peak insulin concentration did not reach the blue I_{on}.

Besides myocytes, adipocytes also respond to insulin for glucose uptake. It is the author's suspicion that they should in general have larger I_{on} values than those

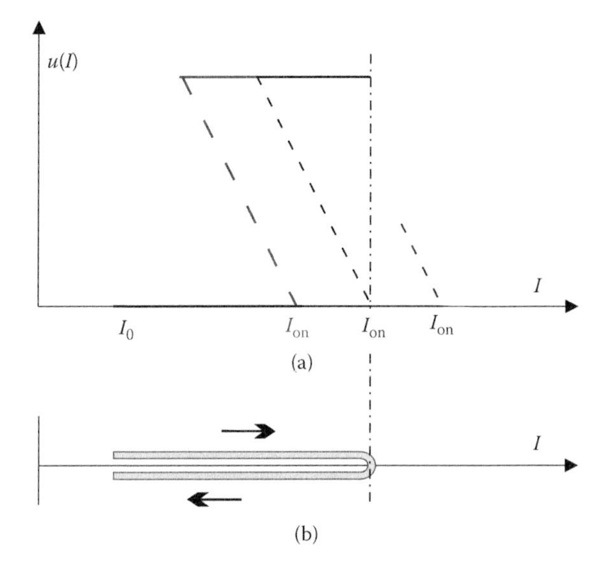

FIGURE 3.4 The I_{on} value may be heterogeneous among the cells. (a) The bistable switch of three cell types. The black switch represents the majority of cells. The red switch represents cells having a smaller I_{on} value. The blue switch represents cells having a larger I_{on} value. (b) The rise and fall of insulin.

of myocytes. The major role of adipose tissues is to store energy; thus there should be no problem that they do not absorb glucose at all. If the person is overeating, for example, then the plasma insulin level begins to exceed the I_{on} values of some adipocytes, which allows the surplus glucose to enter these adipocytes and to convert into fat for storage. Indeed, there is great heterogeneity among adipocytes in terms of insulin responsiveness. It is known that smaller adipocytes are more sensitive to insulin (have smaller I_{on}) than larger adipocytes [35].

3.5 GLOBAL OPTIMALITY

3.5.1 ASSIGNMENT OF PARAMETERS

The local optimality results are independent of the system parameters and are thus general. No matter how the parameter values change, an optimal control must be a square pulse (as a function of time) or a bistable switch (as a function of insulin concentration). The global optimal control, however, is parameter specific. That is, the best control in terms of one set of system parameters is not the best control in terms of another set of parameters. Because the parameters of the glucose-insulin feedback system are variable (at least, the function $m(t)$ is different from meal to meal), there is no global optimality, strictly speaking. Nevertheless, the study of global optimality is still meaningful, because one can use a typical set of parameters that corresponds to normal physiology. For example, one can parameterize the meal function $m(t)$ carefully so that it corresponds to the normal amount of a meal. The globally optimal control thus obtained should be very representative. More importantly, the determination of the globally optimal control may reveal further design principles, as will be demonstrated shortly.

In the following, the parameters of the system are chosen from the literature. They are summarized in Table 3.1.

3.5.2 SEARCH FOR THE GLOBALLY OPTIMAL CONTROL

The globally optimal control is the best among the locally optimal controls. Its determination thus entails the enumeration of all the local optima and the selection of the best. Because these local optima are characterized by only two parameters, t_{on} and t_{off}, the search is only two dimensional — a relatively easy task. In the t_{on} versus t_{off} plane, the area of all the meaningful points (t_{on}, t_{off}) is illustrated in Figure 3.5. There are 8.3×10^5 points in the area, up to a grid size 0.05 min for both axes. For each point, the homeostasis time T and the utilization rate η are calculated in the following way. One first constructs a 0-u_{max}-0 control $u(t)$ that switches on at time t_{on} and switches off at time t_{off}. The control $u(t)$ is then used to run the mathematical model to obtain $G(t)$ and $I(t)$, which apparently follows an exponential decay finally. It is well known that an exponentially decaying state will never exactly reach the steady state (although it approaches the steady state infinitely closer). Therefore, the values $(1+\varepsilon)G_0$ and $(1+\varepsilon)I_0$ ($\varepsilon = 0.1$ is used here) are set as the indicators of homeostasis, instead of G_0 and I_0. In other words, the homeostasis time T is defined as the first time for which $G(t) \le 1.1G_0$ and $I(t) \le 1.1I_0$ hold forever. To obtain

TABLE 3.1
The Values of the Parameters in Equations (2.1, 2.2) and Their Explanations

Parameter	Value	Unit	Remark
G_0	85	$\text{mg}\cdot\text{dl}^{-1}$	The median value of the normal range.
I_0	7.5	$\mu\text{U}\cdot\text{ml}^{-1}$	The median value of the normal range.
s	0.425	$\text{mg}\cdot\text{dl}^{-1}\cdot\text{min}^{-1}$	The value is 0.6 in [3]. The modification is to satisfy the constraint $s/\lambda = G_0$.
λ	0.005	min^{-1}	The value is 0.001 in [3]. The modification is to satisfy the constraint $s/\lambda = G_0$.
k	0.3	min^{-1}	[3] and [32].
G_h	141.4	$\text{mg}\cdot\text{dl}^{-1}$	Obtained from $G_h = \alpha^{0.5}$ and $\alpha = 20{,}000$; [20] and [33].
n	5		$n \geq 4$ is required to fit the clinical data in [30] and [24].
$f_{\max}$	30.91	$\mu\text{U}\cdot\text{ml}^{-1}\cdot\text{min}^{-1}$	Not an independent parameter. Calculated from the relation $f_{\max} = kI_0/\left(1 + (G_h/G_0)^n\right)$.
$u_{\max}$	0.02	min^{-1}	$u_{\max} >> \lambda$.
τ_1	10	min	To mimic the glucose intake pattern.
τ_2	20	min	To mimic the glucose intake pattern.
τ_3	50	min	To mimic the glucose intake pattern.
ξ	0.95		To adjust smoothness of the peak of $m(t)$.
M	1.5	$\text{mg}\cdot\text{dl}^{-1}\cdot\text{min}^{-1}$	The peak rate of exogenous glucose supply. The choice renders the dynamics close to the data in [30] and [24].
η			The utilization rate. The value 0.1 was used to generate Figure 3.2.

the utilization rate, one just uses the formula $\eta = (t_{\text{off}} - t_{\text{on}})/T$, the duty cycle of a 0-$u_{\max}$-0 control. In this way, a dataset is formed containing the T and η values of each point in the area.

The globally optimal control is in terms of a given utilization rate. Therefore, it should be searched for among all the points whose utilization rates are within a narrow range $\eta \pm \Delta\eta$, where the tolerance $\Delta\eta$ (the value 5×10^{-5} is used here) accounts for the fact that no controls have a utilization rate exactly η. The set with $\eta = 0.1$ is first studied, which contains 1048 points whose T values are heterogeneous (Figure 3.6(a)). Figure 3.6(b) shows the corresponding $u(t)$ of three points in the set: the green, blue, and red controls achieve the slowest ($T = 373$ min), an average ($T = 290$ min), and the fastest ($T = 203.8$ min) homeostasis, respectively. The red control is therefore the globally optimal control for $\eta = 0.1$. The homeostasis time of the globally optimal control, $T = 203.8$ min, is realistic because it typically takes about four hours to restore homeostasis after a regular meal [10].

In fact, Figure 3.2 was produced by using the red control in Figure 3.6(b). One sees that $t_{\text{on}} = 70.5$ min happens to be the time when the uncontrolled dynamics $\hat{G}(t)$ and $\hat{I}(t)$ are at their peaks. In other words, the peak of the controlled dynamics

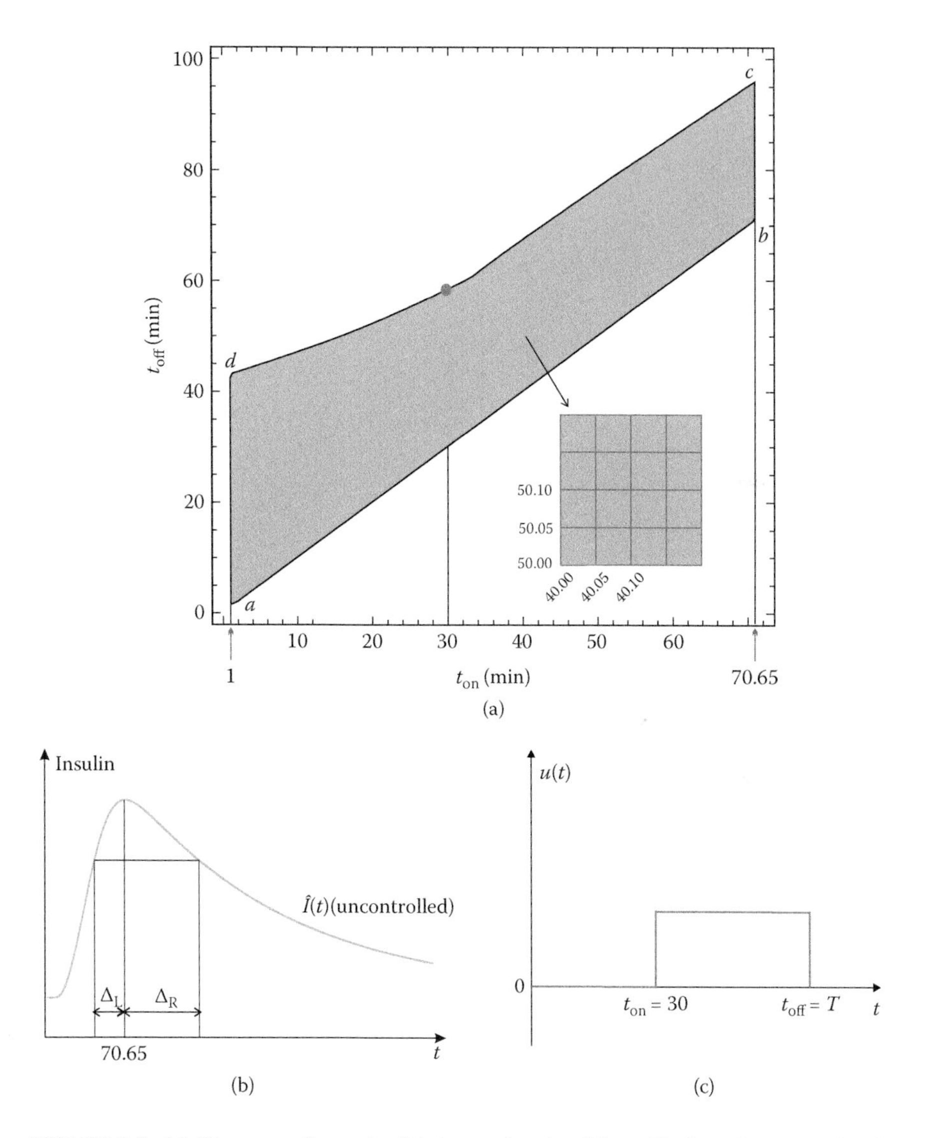

FIGURE 3.5 (a) The area of meaningful (t_{on}, t_{off}) pairs. The grid size of enumeration is 0.05 min for both axes. (b) The uncontrolled insulin level is used to explain the right boundary of the area. (c) A control with $t_{\mathrm{off}} = T$ is used to explain the upper boundary of the area. The control corresponds to the green dot in the area.

is the same with the peak of the uncontrolled dynamics: $G(t_{\mathrm{on}}) = \hat{G}(t_{\mathrm{on}}) = 133$ mg/dl and $I(t_{\mathrm{on}}) = \hat{I}(t_{\mathrm{on}}) = 43.7$ μU/ml, which confers great advantages (see the next section). Because of the relation $I(t_{\mathrm{on}}) = I_{\mathrm{on}}$, the bistable control $u(I)$ has the switch-on threshold $I_{\mathrm{on}} = 43.7$ μU/ml (Fig. 3.6(C)).

Because $\hat{I}(t_{\mathrm{on}}) = 43.7$ μU/ml is the peak of the uncontrolled insulin dynamics (which is not affected by any controls), it is suspected that the globally optimal controls

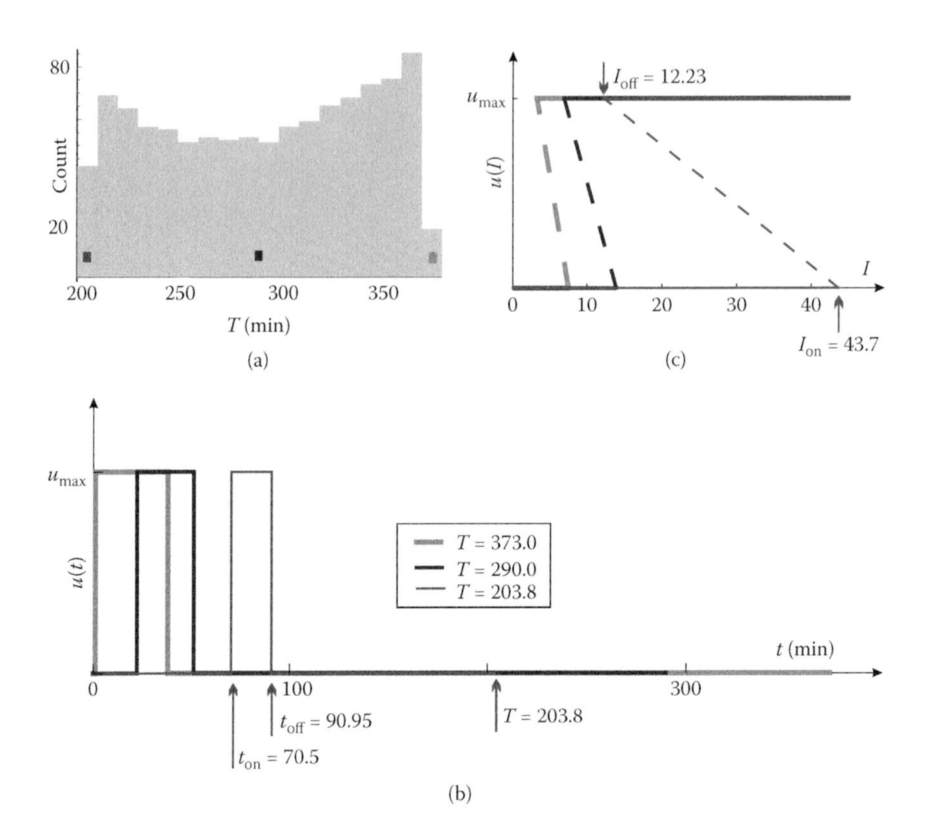

FIGURE 3.6 The set of locally optimal controls whose utilization rates are within the range $\eta \pm \Delta\eta$, where $\eta = 0.1$ and $\Delta\eta = 5 \times 10^{-5}$. (a) The distribution of control counts over the T values. (b) Three $u(t)$ are illustrated: the worst control (green, $T = 373$), an average control (blue, $T = 290$), and the globally optimal control (red, $T = 203.8$). (c) The three controls in the form of $u(I)$. They are obtained by following the conversion procedure in Fig. 3.2. See color insert.

for other small η values may also have I_{on} around 43.7 μU/ml, which is indeed the case. The globally optimal control $u\,(I)$ for 10 different η values 0.01, 0.02, 0.03, 0.04, 0.05, 0.10, 0.15, 0.20, 0.25, and 0.30, are determined; 6 of them are shown in Figure 3.7. The first eight cases all have $I_{on} = 43.7$ μU/ml. Therefore, I_{on} is indeed independent of η as long as η is not too large. For $\eta = 0.25$ and 0.30, I_{on} deviates to 41.8 and 37.7, respectively.

As η increases, I_{off} apparently decreases. The delayed switching off corresponds to a longer duration of insulin action and leads to a shorter homeostasis time. As η increases from 0.05 to 0.30, T decreases from 363.5 to 79.3 min.

3.5.3 Insights from The Obtained Global Optima

Figures 3.2 and 3.7 reveal an interesting constancy: the switch-on threshold I_{on} of the globally optimal controls maintain around 43.7 μU/ml, which corresponds to the peak level of the uncontrolled dynamics $\hat{G}(t)$ and $\hat{I}(t)$. The uncontrolled dynamics

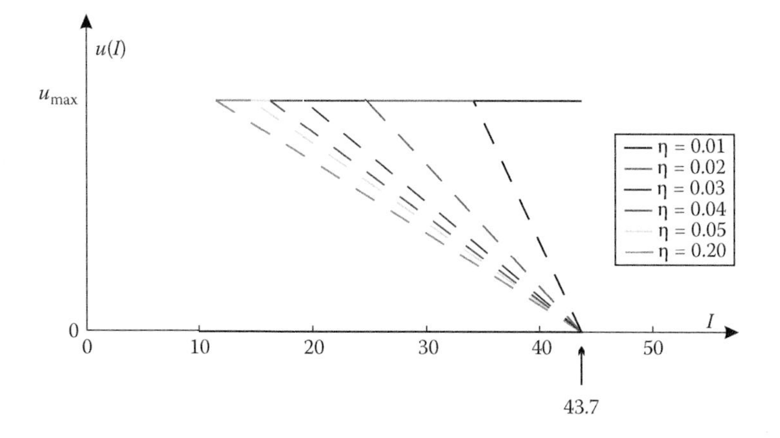

FIGURE 3.7 The globally optimal control $u(I)$ for six η values: 0.01, 0.02, 0.03, 0.04, 0.05, 0.20.

correspond to the (nonexistent) condition that muscles do not uptake glucose, that is, plasma glucose clearance primarily depends on the brain. This design has at least two advantages. The first advantage is rapidity. The peak of $\hat{G}(t)$ corresponds to the largest possible plasma glucose concentration ($G(t_{on}) = 133$ mg/dl). Thus, absorbing glucose around $t_{on} = 70.5$ min would yield the maximum efficiency (because the rate $u_{max}G(t_{on})$ is the largest possible). The second advantage is brain safety. The switch-on threshold $I_{on} = 43.7$ μU/ml is the peak of $\hat{I}(t)$. This large threshold would prevent fortuitous glucose uptake triggered by small insulin levels, thus sparing glucose for the brain. Interestingly, although the globally optimal control was selected according to rapidity, it is also optimal in terms of its opposite — brain safety.

The suboptimal control shown in blue Figure 3.6 is also used for simulating the system dynamics, with the result presented in Figure 3.8. Because the control turns on early ($t_{on} = 22.25$), the spikes of $G(t)$ and $I(t)$ are suppressed. However, this marked early suppression does not bring a long-term benefit: $T = 290 > 203.8$. Moreover, $I_{on} = 13.9$ μU/ml is too small to protect the brain. Indeed, the basal insulin level of a normal person is around $I_{on} = 10$ μU/ml, which is close to the small $I_{on} = 13.9$ μU/ml. The poor performance of the blue control clearly demonstrates the superiority of the globally optimal control.

The bistable switch works in a similar way as a valve controlling fluid power. During the fasting state, the valves are closed firmly so that the limited glucose all goes to more important organs, especially the brain. The opening of the valves is significantly delayed to wait until the plasma glucose concentration builds up (so that the subsequent glucose transportation will be more efficient). Retardation is good in certain situations! Following the rise of glucose, the insulin concentration also rises and finally exceeds a preset threshold I_{on}. The tissue cells can then enjoy the surplus glucose. The valves close and close fully when the insulin level drops below another threshold I_{off}, which is smaller than I_{on}. The valve is thus hysteretic.

In the search for the global optimum, we have assumed that the cells have the same I_{on} value (the homogeneity assumption), which is apparently a simplification.

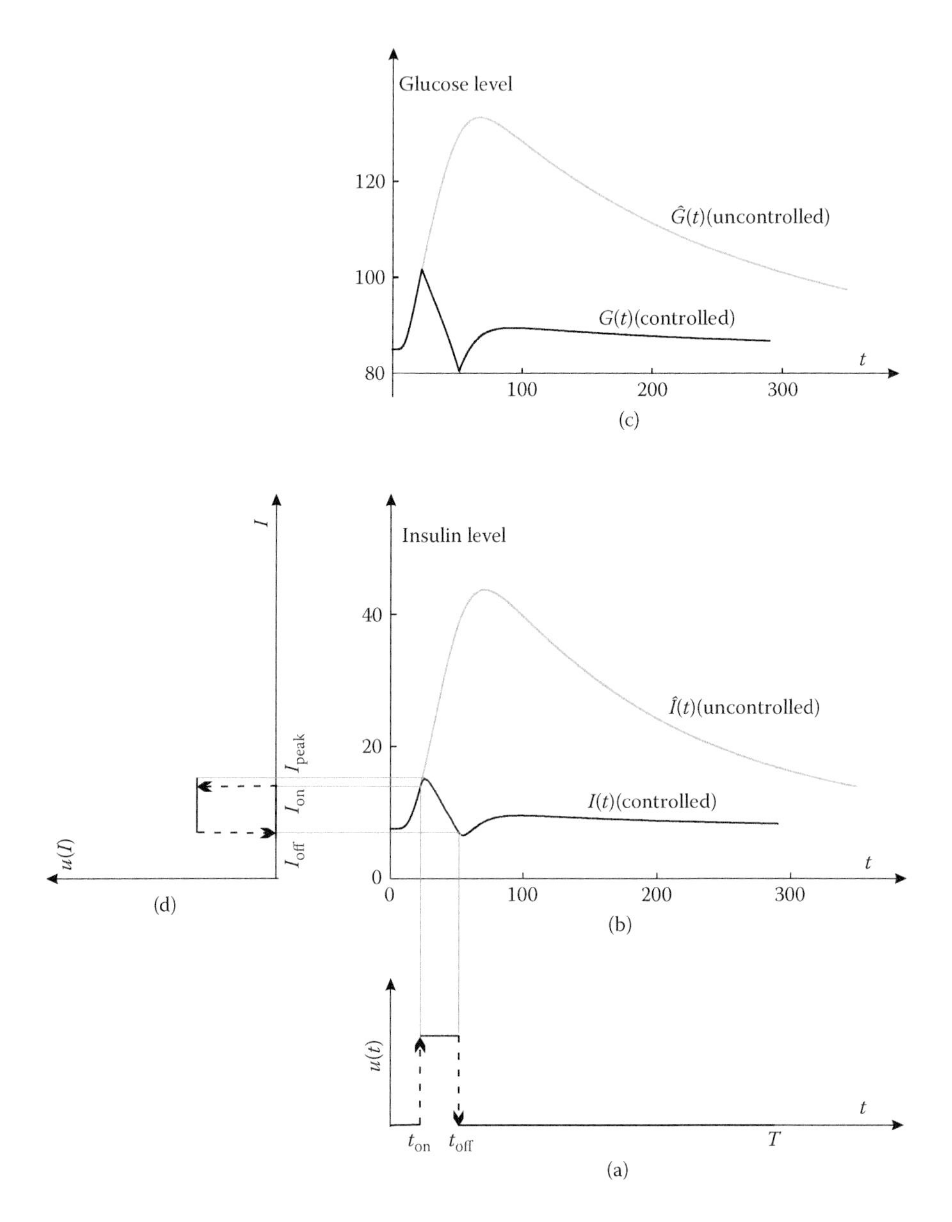

FIGURE 3.8 A computer simulation of the glucose-insulin system. (a) The control $u(t)$ is taken from the blue control in Figure 3.6(b). (b) The controlled and uncontrolled insulin dynamics. (c) The controlled and uncontrolled glucose dynamics. (d) The control in the form of $u(I)$.

Yet proper simplifications are sometimes crucial to reveal mechanisms — a good example is the point mass in physics. Although mass does not really exist as a point, the assumption is crucial to solve many practical problems and has become general. The homogeneity assumption has rendered the discovery of the biphasic regulation of glucose homeostasis conferred by the bistable insulin response. Moreover, the global optimality result suggests that it is advantageous for the majority of myocytes to have a large switch-on threshold I_{on}, as large as the peak of the uncontrolled insulin dynamics. Such a long delay of insulin action is counterintuitive in the first place, yet it is the only way in which conflicting requirements can be reconciled. It should be noted that in reality the I_{on} values are heterogeneous. With that being said, the cells' I_{on} values should follow a very sharp normal distribution centering on 40 μU/ml, should the system be "designed" optimally. A related phenomenon, which sounds even more counterintuitive, is that massive glucose uptake takes place primarily during the fall of insulin, for the majority of tissue cells. By intuition one would expect that glucose uptake speeds up during insulin increase and lets up during insulin decrease.

3.5.4 The Yin-Yang Insulin

The biphasic regulation discovered in this book forces us to reconsider the roles played by insulin. Insulin is a peptide hormone produced by beta cells of the pancreas. It has long been known as a growth factor promoting glucose metabolism — the Yang side of insulin action. Since muscles and adipose tissues constitute the majority of our body mass, the importance of insulin in facilitating glucose utilization is indisputable. Its importance has been highlighted as a consequence of the global epidemic of diabetes — it was projected that about one-third of children born after the year 2000 would develop diabetes at some stage of their lives. Type 2 diabetes is caused by insulin resistance; tissue cells do not respond to insulin sufficiently to elicit glucose uptake.

The emphasis on the Yang side of insulin inevitably overshadows its Yin side, which, in the author's opinion, is even more important. For most cell types, glucose is just an alternative fuel. Muscles do not have any urgency in obtaining glucose — their preferred fuel is FFA. On the other hand, cells intrinsically do not have problems in getting glucose, due to a variety of GLUTs expressed in the cell. Neurons have GLUT3 and can thus take up glucose efficiently even when the plasma glucose concentration is very low. Unicellular organisms (such as budding yeast) can readily take up nutrients in the environment, without help from additional hormones. Muscles, if their dependence on glucose were indeed critical, would have evolved mechanisms to secure their glucose utilization (such as GLUT3 expression or GLUT1 overexpression). Then, what are the purposes of developing insulin as a *growth* factor?

The answer to that question would highlight the importance of the Yin side of insulin. The essential role of insulin is not so much promoting growth, but *delaying* glucose metabolism of certain tissues that are relatively unimportant. Compared with other organs (brain, liver, pancreas, etc.), skeletal muscles and adipose tissues are indeed less important. Loss of muscle or fat, even in large quantity, is usually not life threatening. As a matter of fact, the loss of fat makes obese people even healthier. The loss of muscle is a natural process. After about the age of 25, there is a 0.5% to 1% loss of muscle per year for an average person. The I_{on} value of myocytes and adipocytes thus sets a bar to delay (or to block) their glucose uptake so that glucose can be first utilized by more important cells such as neurons, whose glucose uptake is insulin independent. This is especially important if the plasma glucose concentration is small, because the muscles (which are so massive!) certainly constitute a menace to the brain if their glucose uptake is not controlled. If the plasma glucose is so abundant that the plasma insulin level exceeds I_{on}, then myocytes can enjoy glucose. As luck would have it, there is still no competition, due to the abundance of glucose. This is exactly the beauty of the design.

Bistability, characterized by all-or-none and hysteresis, is a perfect design to render the Yin-Yang action of insulin. This design has worked well throughout evolution and became problematic only recently. In the evolutionary time scale, abundance of food and sedentary lifestyle are only recent things, which constitute acute environmental changes that may challenge the evolved optimality that worked well for our ancestors. The twin epidemics of obesity and diabetes have raised an alarm about maladaptation. It will be shown in Chapter 7 that bistability can be severely deformed by modern factors and become less optimal or even detrimental in some situations. The deformation of bistability may partly explain the simultaneous escalation of epidemics of cancer, obesity, and diabetes. Stay tuned.

4 Bistability as a Fundamental Phenomenon

Bistability is a ubiquitous dynamical behavior. Instead of a graded response to the stimulus, a bistable molecule toggles between active and inactive states (all-or-none), but the threshold concentration of the stimulus for activation significantly differs from that for deactivation (hysteresis). Biological examples of bistability include the *Escherichia coli* lac operon, mitogen-activated protein kinase cascades, cell cycle circuits, the insulin signaling pathway, and synthetic gene switches.

Bistability corresponds to multiple fine properties: it produces a "memory" of a transient stimulus, which is important for cell differentiation and cell cycle progression; it enables cell's robustness to noises while responding sensitively to signals; it is a mechanism through which small noncoding RNAs can mediate gene regulation; it confers adaptivity by adjusting hysteresis; it helps achieve biological rhythms with widely tunable frequency and near-constant amplitude. On the other hand, disturbances to bistability often lead to complex diseases such as cancer and diabetes.

However, these fine properties do not explain why bistability evolved in organisms. First, the wide variety of miscellaneous properties appears too special to explain ubiquity. Occam's razor (the principle that plurality should not be postulated unless absolutely necessary) argues for a simple and general explanation. Second, bistability may not be unique in conferring these properties. Other dynamics can possibly realize the same physiological properties as well.

The first concern can be addressed by seeking an overarching principle that involves only general notions. The second concern can be addressed by demonstrating that the overarching principle necessitates bistability. One starts from the principle and then deduces how the organism should be "designed." If bistability emerges as an indispensable mechanism, then bistability proves to be necessary for the overarching principle, and the generality of the principle can explain the ubiquity of bistability.

Optimal homeostasis is such an overarching principle. Darwin's theory of natural selection provided an obvious mechanism for optimization in biology: more optimized individuals are more likely to survive. After many rounds of natural selection, a species is optimally adapted to the environment it evolved in. Homeostasis, first defined by Claude Bernard and later by Walter Bradford Cannon, has become a central concept in biology. Primarily through negative feedback control, homeostasis allows an organism to maintain a stable parameter such as temperature. The negative feedback control must accommodate competing or even conflicting internal requirements and possibly wide environmental variations. Those that achieve optimal compromise have an advantage in surviving natural selection. Optimal homeostatic regulation appears to be an overarching principle acquired through evolution.

We have reached the conclusion that optimal glucose homeostasis necessitates bistable control. Can this conclusion be extended to a general homeostatic system, not merely glucose homeostasis? One important fact of the last chapter is that the bistability solution is largely independent of the values of the parameters in the mathematical model. The discussion of local optimality, namely, the square pulse solution of the optimal $u(t)$, was obtained by symbolic computation only (no numerical computation was involved). The determination of the optimal $u(I)$ needs some numerical computation — one needs to synthesize $u(I)$ from $u(t)$ and $I(t)$; and the insulin dynamics $I(t)$ have to be calculated numerically. Nevertheless, the solution to $u(I)$ is almost always a bistable switch. If one uses some bizarre set of parameter values, it is possible that the optimal solution is not bistability. But such bizarre parameters apparently have no practical correspondence. Therefore, the mathematical model (Equations (2.1) and (2.2)) represents a general negative feedback system, and bistability is thus a fundamental control mechanism in nature.

To better demonstrate the generality of Equations (2.1) and (2.2), we can start from a simple model:

$$\frac{dG}{dt} = 1 - G, \tag{4.1}$$

which stabilizes at $G_0 = 1$. With a transient stimulation $m(t)$ added to the right-hand side of Equation (4.1), the state point leaves G_0 due to the stimulation. However, the state point will finally return to G_0, because $m(t)$ is transient ($m(t) \to 0$ as $t \to 0$). If the return is slow, one can expedite the return by adding a control term $-u$, which turns the model into

$$\frac{dG}{dt} = 1 - G + m(t) - uG. \tag{4.2}$$

If u is a constant, then the control is open-loop (non-feedback). To be a feedback control, u must change adaptively with a signal reflecting the level of G. The signal can be G itself. For example, $u = \gamma G$ is a proportional control that leads to the dynamics

$$\frac{dG}{dt} = 1 - G + m(t) - \gamma G^2. \tag{4.3}$$

However, it is possible that G is inconvenient or expensive for signal transduction. In this event, another entity I, which tracks the change of G (see Equation (4.5)), serves as the feedback signal. For example, the small molecule glucose is less suitable for signaling when compared with insulin, a peptide that has a sufficiently complex three-dimensional structure. The two-component model can be described by

$$\frac{dG}{dt} = 1 - G + m(t) - u(I)G, \tag{4.4}$$

$$\frac{dI}{dt} = G - I. \tag{4.5}$$

Being negative, the control $-u(I)$ is a negative feedback one.

Equations (4.4) and (4.5) represent a general homeostatic system achieved by negative feedback. This general model, however, is only a special case of Equations (2.1) and (2.2) (with $s = \lambda = k = 1$, and $f(G) \equiv G$), which implies that Equations (2.1) and (2.2) are even more general.

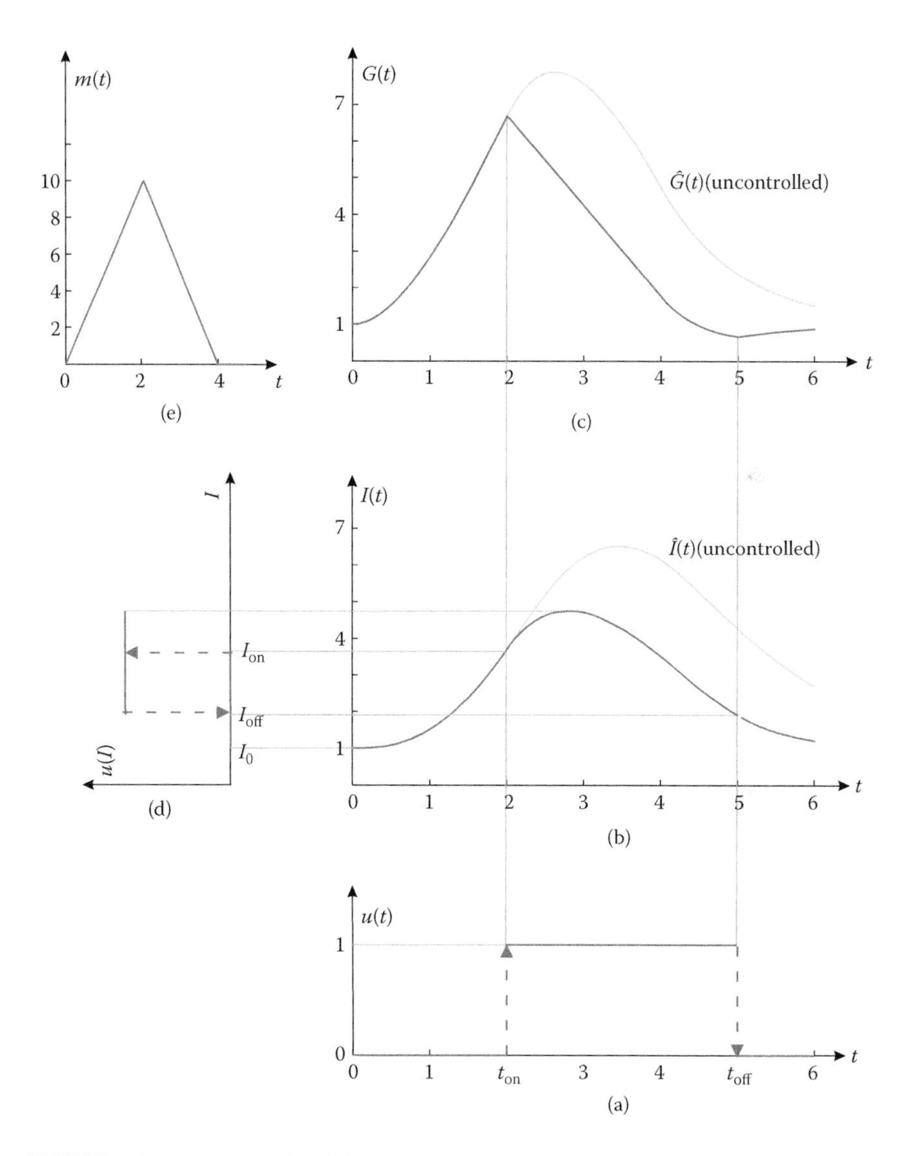

FIGURE 4.1 A computer simulation of Equations (4.4) and (4.5). (a) The control $u(t)$ with $u_{\max} = 1$, $t_{\text{on}} = 2$, and $t_{\text{off}} = 5$. (b) The controlled (black) and uncontrolled (grey) insulin dynamics. (c) The controlled and uncontrolled glucose dynamics. (d) The control in the form of $u(I)$. (e) The function $m(t)$ is a triangular function centering on $t = 2$ and with height 10.

The analysis in Equations (2.1) and (2.2) has shown that locally optimal $u(t)$ must be 0-$u_{\max}$-0 (Equation (3.2)), which certainly applies to the model described by Equations (4.4) and (4.5), which is more specialized. An intuitive explanation of this result is as follows.

Without constraints, it is easy to see that the optimal control is "all" (i.e., $u(t) \equiv u_{\max}$). By consistently applying the maximum force $u_{\max}$, the homeostasis time is certainly minimal. But this "all" solution might be unacceptable owing to many biological constraints: a great deal of energy would be wasted and the constitutive activation of a molecule might lead to problems such as cancer, potential conflicts with other physiological requirements, and so on.

With the constraint $\eta < 1$, the optimal control cannot be "all." Analysis has revealed that it is still simple: "all-or-none." The control is a square pulse: turning on fully at time t_{on} and turning off fully at time t_{off}. This result implies that a bell-shaped $u(t)$ (increases gradually and decreases gradually) cannot be an optimal control.

The optimal $u(t)$, after conversion into $u(I)$, should be a bistable switch. In Figure 3.2, an example was shown in the context of the glucose-insulin feedback system. To demonstrate that the result was not specific to glucose homeostasis nor obtained by chance, a $u(t)$ is generated with $t_{on} = 2$ and $t_{off} = 5$ (Figure 4.1(a)). With this $u(t)$, Equations (4.4) and (4.5) are integrated to obtain $G(t)$ and $I(t)$, which are shown in Figures 4.1(c) and 4.1(b), respectively. The functions $u(t)$ and $I(t)$ are synthesized to obtain $u(I)$, which is again a bistable switch (Figure 4.1(d)). Note that the parameters here ($s = \lambda = k = 1$) differ significantly from those of the glucose-insulin system ($s = 0.425$, $\lambda = 0.005$, and $k = 0.3$). Here the function $m(t)$ is a triangular function (Figure 4.1(e)), which is different from the one used for the glucose-insulin system (Figure 2.1). Here $f(G)$ is just the linear function G, while in the glucose-insulin system $f(G)$ is a nonlinear Hill function (Figure 2.2). Despite all these differences, the same bistability feature was obtained. This is in line with the theoretical results that the optimal control is largely independent of the values of the parameters in Equations (2.1) and (2.2).

The ubiquity of bistability now has a reason. As organisms become more complex, homeostatic regulation of the internal environment becomes more difficult owing to the increasing number of competing requirements that have to be accommodated simultaneously. It has been shown that bistability is necessary to confer optimal compromise in homeostatic regulation.

Part II

Molecular Level: What Can Be Provided?

Part I of this book describes phenomena that occur primarily at the organismal level, the basic elements of food intake and energy metabolism, which are of vital importance to our everyday life. Even though our focus was limited to glucose metabolism, we already found great complexities in regulating the glucose-insulin feedback system. The complexities primarily arise from the fact that different organs have different priorities in utilizing glucose, and thus a mechanism must have evolved to reconcile conflicting requirements from different organs and to satisfy other constraining conditions. In Part I, bistability was identified as a *necessary* control mechanism in regulating glucose-insulin homeostasis. That is, a myocyte is required to respond to insulin in a bistable manner in order to satisfy all the physiological constraints. However, nothing has been said thus far as to how the bistable control $u(I)$ can be realized within a cell.

The optimal controller is generated by a molecular circuit within almost every cell in the human body, in much the same way as a controller in engineering can be generated by an electrical or electronic circuit. The circuit is known as the PI3K-AKT-TOR pathway and is described in detail in Chapter 7. In general, a molecular circuit is a network of interacting biomolecules in a cell: RNAs, DNAs, proteins, etc. Moreover, information processing in a biomolecular network is carried out primarily by chemical reactions. A living cell has receptors on its surface, which, in response to a specific chemical in the environment, generates a signal that is relayed by subsequent signaling molecules. Besides linear information propagation, a biomolecular network may have various modules as necessary components to complete functional output. Some modules are recurring motifs because they have certain structural characteristics (e.g., feedback and feedforward loops) and dynamical patterns (e.g., switch and oscillation). These network components work in unison to provide biological functions.

Therefore, Part II of this book primarily focuses on the molecular level to see how the biomolecules are organized and how their combined actions can achieve optimal functional outputs. Aspects of biochemical reactions (enzyme kinetics, sensitivity, etc.) will first be introduced, followed by an introduction to structural and functional properties of biomolecular networks. We will then zoom in to study the PI3K-AKT-TOR pathway, the intracellular controller of glucose homeostasis. The pathway also has other important functions and they will be described.

5 Molecular Biology

5.1 CENTRAL DOGMA OF MOLECULAR BIOLOGY

The central dogma of molecular biology describes the flow of genetic information within a biological system. As Francis Crick stated in 1970: "The central dogma of molecular biology deals with the detailed residue-by-residue transfer of sequential information. It states that such information cannot be transferred back from protein to either protein or nucleic acid" [8]. Francis Crick later acknowledged that the word "dogma" was not accurate. Nevertheless, the significance and usefulness of the central dogma make it well known in biology.

Sequential information is carried by three major classes of biopolymers: DNA, RNA, and protein. There are therefore $3^2 = 9$ conceivable direct transfers of information that can occur between the three classes. Among the nine, there are three *unknown information transfers*

$$\begin{aligned}&\text{protein} \to \text{DNA}\\&\text{protein} \to \text{RNA}\\&\text{protein} \to \text{protein}\end{aligned}$$

which are believed never to occur; three *special information transfers*

$$\begin{aligned}&\text{RNA} \to \text{DNA}\\&\text{RNA} \to \text{RNA}\\&\text{DNA} \to \text{protein}\end{aligned}$$

which can occur, but only under specific conditions in the case of some viruses or in a laboratory; and three *general information transfers*

$$\begin{aligned}&\text{DNA} \to \text{DNA}\\&\text{DNA} \to \text{RNA}\\&\text{RNA} \to \text{protein}\end{aligned}$$

which are believed to occur normally in most cells.

The general transfers describe the normal flow of biological information: DNA can be copied to DNA (DNA replication), DNA information can be copied into mRNA (transcription), and proteins can be synthesized using the information in messenger RNAc (mRNA) as a template (translation).

5.1.1 DNA Replication (DNA $\to$ DNA)

As the final step in the central dogma, DNA replication is the process by which genetic information can be faithfully transmitted to the progeny of any cell or organism. Replication is carried out by the replisome, which consists of a helicase that unwinds the superhelix as well as the double-stranded DNA helix, and DNA polymerase and its associated proteins, which insert new nucleic material in a sequence-specific manner. This process typically takes place during the S phase of the cell cycle.

5.1.2 Transcription (DNA → RNA)

This is the process by which the information contained in a section of DNA is transferred to a newly assembled piece of mRNA. It is facilitated by RNA polymerase and transcription factors. In eukaryotic cells the primary transcript (pre-mRNA) must be processed further in order to ensure translation. This normally includes a 5′ cap, a poly-A tail, and splicing. Alternative splicing can also occur, which contributes to the diversity of proteins any single mRNA can produce.

5.1.3 Translation (RNA → Protein)

The mature mRNA eventually finds its way to a ribosome, where it is translated into protein. In eukaryotic cells, translation and transcription are usually separate, with translation occurring in the cytoplasm and transcription occurring in the nucleus. Therefore, mRNA must be transported out of the nucleus into the cytoplasm, where it can be bound by ribosomes. The mRNA is read by the ribosome as triplet codons, usually beginning with an AUG (adenine-uracil-guanine), or initiator methionine codon downstream of the ribosome binding site. Complexes of initiation factors and elongation factors bring aminoacylated transfer RNAs (tRNAs) into the ribosome-mRNA complex, matching the codon in the mRNA to the anticodon on the tRNA, thereby adding the correct amino acid in the sequence encoding the gene. As the amino acids are linked into the growing peptide chain, they begin folding into the correct conformation. Translation ends with a UAA, UGA, or UAG stop codon. The nascent polypeptide chain is then released from the ribosome as a mature protein. In some cases the new polypeptide chain requires additional processing to make a mature protein. The correct folding process is quite complex and may require other proteins, called chaperone proteins.

5.2 POSTTRANSLATIONAL MODIFICATION

In addition to the information flow described by the central dogma, there is another kind of information transfer known as posttranslational modification (PTM) — chemical modification of a protein after its translation. As we know, translation is the process of adding amino acids in sequence to form a polypeptide (protein). The kind of amino acids added (from the 20 choices) is completely determined by the mRNA information. After translation, the amino acids can be modified in many different ways (see below) so that the protein's functions can be extended. There are four major categories of PTM.

5.2.1 Protein Structural Changes

- Disulfide bridge, which is the covalent linkage of two cysteine amino acids. The linkage is also called an SS-bond.

- Proteolytic cleavage, which is the cleavage of a protein by a protease at a peptide bond. A protease is any enzyme that conducts proteolysis.
- Racemization of proline by prolyl isomerase. Racemization refers to the conversion of an enantiomerically pure mixture (one where only one enantiomer is present) into a mixture where more than one of the enantiomers are present.

5.2.2 Amino Acid Chemical Changes

- Citrullination (deimination), which is the conversion of arginine into citrulline.
- Deamidation, which converts glutamine into glutamic acid and converts asparagine into aspartic acid.
- Eliminylation, which is the conversion to an alkene by beta-elimination of phosphothreonine and phosphoserine, or dehydration of threonine and serine, as well as by decarboxylation of cysteine.
- Carbamylation, which is the conversion of lysine into homocitrulline.

5.2.3 Addition of Other Proteins or Peptides

- ISGylation, the covalent linkage to the ISG15 protein (Interferon Stimulated Gene 15).
- SUMOylation, the covalent linkage to the SUMO protein (Small Ubiquitin-related MOdifier).
- Ubiquitination, the covalent linkage to the protein ubiquitin.
- Neddylation, the process by which the ubiquitin-like protein Nedd8 is conjugated to its target proteins.
- Pupylation, the process by which the pup protein is conjugated to the amino groups of lysine amino acids in target proteins.

5.2.4 Addition of Functional Groups

There are many different kinds of functional groups that can be used to modify a protein. For example, hydrophobic groups can be added to a protein to confer cell membrane targeting and localization; cofactors can be added to a protein kinase to enhance its enzymatic activity; nonenzymatic additions are also involved both in vivo and in vitro. The addition of smaller chemical groups is even more versatile. There are many important small modifications, including but not limited to:

- Hydroxylation, which is the addition of a hydroxyl group (-OH). The principal residue to be hydroxylated in proteins is proline.
- Acetylation, which is the addition of an acetyl group, either at the N-terminus of the protein or at lysine residues. The reverse process is called deacetylation. A well-known example is histone acetylation and deacetylation — a mode of epigenetic control.
- Methylation, which is the addition of a methyl group, usually to lysine or arginine residues in the sequence of a protein. The reverse process is called demethylation.

- Phosphorylation, which is the addition of a phosphate group, usually to serine, threonine, tyrosine, or histidine residues. The reverse process is called dephosphorylation. Phosphorylation and dephosphorylation are ubiquitous processes in biochemical reactions. The following section is thus devoted to a detailed description of phosphorylation and dephosphorylation.

5.3 PHOSPHORYLATION AND DEPHOSPHORYLATION

Phosphorylation is the addition of a phosphate (PO_4^{3-}) group to a protein or other organic molecule, thereby altering their function and activity. The enzymes catalyzing phosphorylation are often called kinases. Dephosphorylation, as the name suggests, is the removal of the phosphate group. The enzymes catalyzing dephosphorylation are often called phosphatases. Phosphorylation and dephosphorylation, which turn many protein enzymes on and off, play a particularly significant role in a wide range of cellular processes. As of January 2013, the Medline database returns more than 210,000 articles on the subject, largely on protein phosphorylation and dephosphorylation.

5.3.1 Sites of Protein Phosphorylation

Protein phosphorylation usually occurs on serine, threonine, histidine, and tyrosine residues in eukaryotes. Phosphorylation on serine is the most common, followed by threonine, and then histidine. Tyrosine phosphorylation is relatively rare. However, since tyrosine phosphorylated proteins are relatively easy to purify using antibodies, tyrosine phosphorylation sites are relatively well understood.

The number of distinct sites of phosphorylation in a given cell is actually innumerable. First, there are thousands of different kinds of proteins in any particular cell and it is estimated that as many as half of them are phosphorylated. Second, a given protein often has multiple distinct sites for phosphorylation. For example, the protein kinase AKT has two common sites of phosphorylation; one is serine-473 (S473) and the other is threonine-308 (T308). TP53 is a tumor suppressor protein that contains more than 18 different phosphorylation sites.

5.3.2 Functions of Protein Phosphorylation

The phosphorylation and dephosphorylation of a protein represent an important regulatory mechanism that occurs in both prokaryotic and eukaryotic organisms. Reversible phosphorylation results in a conformational change in the structure of many enzymes and receptors, causing them to become activated or deactivated. When phosphorylation causes activation, dephosphorylation causes inhibition. For some cases, dephosphorylation causes activation and phosphorylation causes inhibition. For example, the phosphorylation of GSK-3 (by the protein kinase AKT) leads to the inhibition of the activity of GSK-3, instead of activation.

Phosphorylation plays the following regulatory roles:

- Biological thermodynamics of energy-requiring reactions. For example, homeostasis of the body's water content can be maintained by phosphorylation of Na^+/K^+-ATPase during the transport of sodium (Na^+) and potassium (K^+) ions across the cell membrane in osmoregulation.

- Enzyme inhibition. For example, the protein kinase AKT can inhibit, through phosphorylation, many downstream targets, including GSK-3, tuberous sclerosis complex (TSC), and forkhead box protein O1 (FOXO1).
- Protein-protein interaction via recognition domains. Phosphorylation of the cytosolic components of NADPH oxidase, a large membrane-bound, multiprotein enzyme present in phagocytic cells, plays an important role in the regulation of protein-protein interactions in the enzyme.
- Protein degradation. In the late 1990s, it was recognized that phosphorylation of some proteins causes them to be degraded by the ATP-dependent ubiquitin/proteasome pathway. These target proteins become substrates for particular E3 ubiquitin ligases only when they are phosphorylated.

5.3.3 Networks of Phosphorylation

Although other kinds of PTMs exist as forms of molecular interaction, signaling transduction in a biomolecular network is usually dominated by protein phosphorylation and dephosphorylation. Global approaches such as phosphoproteomics, the study of phosphorylated proteins, which is a sub-branch of proteomics, combined with mass spectrometry-based proteomics, have been utilized to identify and quantify dynamic changes in phosphorylated proteins over time. These techniques are becoming increasingly important for the systematic analysis of complex phosphorylation networks. They have been successfully used to identify dynamic changes in the phosphorylation status of more than 6000 sites after stimulation with epidermal growth factor. Another approach to understanding phosphorylation networks is by measuring the genetic interactions between multiple phosphorylating proteins and their targets. This helps to reveal interesting recurring patterns of interactions — network motifs.

5.4 ENZYME KINETICS

The mainstream flow of information in molecular biology can be roughly and briefly described as: DNA makes RNA, which then makes protein, which is subject to modification such as phosphorylation. What has not been discussed is how the information transfer is realized.

These biological changes are realized by chemical reactions. Change is the essence of life and gives rise to its complexity. Numerous biochemical reactions occur at every moment in our body which control the flow of energy and information through metabolism and the interconversions among the four main classes of biomolecules: carbohydrates, lipids, proteins, and nucleic acids. Although there are numerous biochemical reactions, they fall into only a few types, such as oxidation and reduction, movement of functional groups within or between molecules, addition and removal of water, and bond-breaking reactions. Biochemical reactions follow all the laws of chemistry, but are generally very fast and more efficient than reactions done in the laboratory, due to many enzymes that catalyze reactions.

Enzymes are remarkable biomolecules catalyzing biochemical reactions. Nearly all known enzymes are protein molecules. In 1903, the French physical chemist Victor

Henri found that enzyme reactions were initiated by a bond between the enzyme and its target molecule — the substrate. The most striking characteristics of enzymes are their catalytic power and specificity. Enzymes can accelerate reactions by many orders of magnitude, even as large as 10^{17}, as demonstrated by the enzyme orotidine 5′-phosphate decarboxylase (OMP decarboxylase) that catalyzes the decarboxylation of orotidine monophosphate (OMP) to form uridine monophosphate (UMP). The enzyme can catalyze the reaction in 18 milliseconds, without which the reaction would take 78 million years to complete [27]. Enzymes are highly specific to the substrate. The specificity is due to the intricate three-dimensional structure of the enzyme, which allows it to interact with the substrate precisely.

The substrates bind to an enzyme's active site and are transformed into products through a series of steps known as the enzymatic mechanism. When the substrate concentration is relatively low, the active sites of the enzyme molecules are largely unoccupied; an increase in the substrate concentration would increase the rate at which the enzyme and substrate molecules encounter each other. In this case, the reaction rate increases linearly with substrate concentration. When the substrate concentration is high, the enzyme active sites are almost all occupied; the reaction rate asymptotically approaches the theoretical maximum. The reaction rate is determined by the intrinsic turnover rate of the enzyme.

In 1913, German biochemist Leonor Michaelis and Canadian physician Maud Menten proposed a mathematical model to describe enzyme kinetics, which soon became a standard model, known as the Michaelis–Menten kinetics [22]. It involves an enzyme binding to a substrate to form a complex, which in turn is converted into a product and the enzyme:

$$E + X \underset{k_{\text{off}}}{\overset{k_{\text{on}}}{\rightleftharpoons}} EX \xrightarrow{k_{\text{cat}}} E + P \tag{5.1}$$

where

X = the substrate
E = the enzyme
EX = the enzyme-substrate compound
P = the product
k_{on} = the rate of the substrate binding with the enzyme
k_{off} = the rate of the substrate leaving the enzyme
k_{cat} = the rate of the intermediate compound EX converting to the product

In the context of protein phosphorylation, X represents the protein substrate, E represents the kinase (enzyme), and P represents the phosphorylated protein. Glucose transportation can also be regarded as an enzymatic reaction, for which case X may represent glucose molecules in the blood; E may represent GLUT, namely, the glucose transporter; EX may represent the GLUT loaded with glucose; and P may represent the same glucose molecules that are now within the cell.

In 1925, British botanist G. E. Briggs and British geneticist J. B. S. Haldane proposed a quasi–steady-state approximation, namely, the assumption that the concentration of the intermediate complex does not change on the time-scale of product formation ($[EX]$ = const) [4]. By this approximation, the velocity of reaction v is determined to be

$$v = \frac{d[P]}{dt} = \frac{V_{\max}[X]}{K_m + [X]}, \tag{5.2}$$

where

$$V_{\max} = k_{\text{cat}}[E_0]$$

is the maximum reaction velocity, which quantifies the capacity of the enzyme in converting the substrate into the product; $[E_0]$ is the enzyme concentration; and

$$K_m = \frac{k_{\text{off}} + k_{\text{cat}}}{k_{\text{on}}}$$

is called the Michaelis constant, which equals the substrate concentration $[X]$ at which the reaction rate is at half-maximum ($v = V_{\max}/2$). A small K_m implies that the enzyme has a high affinity with the substrate, because the reaction rate approaches $V_{\max}$ more quickly (see Figure 5.1(a)). The value of K_m is dependent on the structures of both the enzyme and the substrate, as well as conditions such as temperature and pH.

The constant

$$\frac{k_{\text{cat}}}{K_m}$$

is used to measure the efficiency of an enzyme. It has a theoretical upper limit of 10^8–10^{10} $(\text{M}\cdot\text{s})^{-1}$. A large k_{cat}/K_m implies high capacity (k_{cat} large) and/or high affinity (K_m small). For example, the enzyme fumarase is considered to be superefficient, because its k_{cat}/K_m value is 1.6×10^8. Among all the glucose transporters, GLUT3 has the highest k_{cat}/K_m (which is about 15-fold higher than that of GLUT4), which explains why GLUT3 is abundantly expressed in neurons — the continuous and great demand for glucose of the brain necessitates glucose transportation with the highest efficiency.

The kinetics represented by Equation (5.2) and Figure 5.1 are known as the Michaelis–Menten kinetics.

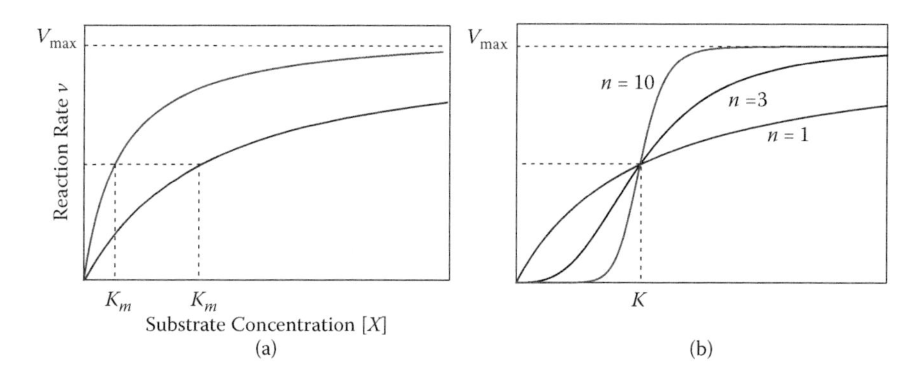

FIGURE 5.1 Enzyme kinetics. (a) The Michaelis–Menten kinetics described by Equation (5.2). The smaller the K_m, the more rapid the response. (b) The increase of sensitivity as the Hill coefficient n increases. The larger the n, the more switch-like the curve.

5.5 ULTRASENSITIVITY

In phenomena such as sensing, and in the regulation of metabolism, it is important that the activation and deactivation of a biomolecule is sensitive to relatively small changes in effector concentration. That is, the response should be switch-like: unresponsive to a subthreshold stimulus but fully responsive once the threshold is exceeded. This dynamical property is known as ultrasensitivity in molecular biology.

Ultrasensitivity can be realized by several mechanisms. These include buffering mechanisms (e.g., decoy phosphorylation sites), stoichiometric inhibitors, changes in localization (e.g., translocation across the nuclear envelope), cooperative binding, and zero-order ultrasensitivity (also known as enzyme saturation). In the following, the last two mechanisms are described in detail, due to their generality.

5.5.1 COOPERATIVE BINDING

In biochemistry, a macromolecule exhibits cooperative binding if its affinity for its ligand changes with the amount of ligand already bound. Cooperative binding is a special case of allostery. Cooperative binding requires that the macromolecule have more than one binding site, since cooperativity results from the interactions between binding sites. If the binding of the ligand at one site increases the affinity for the ligand at another site, the macromolecule exhibits positive cooperativity. Conversely, if the binding of the ligand at one site lowers the affinity for the ligand at another site, the protein exhibits negative cooperativity. If the ligand binds at each site independently, the binding is noncooperative.

Binding cooperativity can be characterized by a number called the Hill coefficient, which is usually denoted by n. The cooperative kinetics are characterized by the Hill function:

$$v = \frac{[X]^n}{K^n + [X]^n}, \tag{5.3}$$

where

v = the fraction of occupied sites where the ligand can bind to the active site of the receptor protein.
$[X]$ = the free (unbound) ligand concentration.
K = the microscopic dissociation constant, which equals the ligand concentration producing half occupation (ligand concentration occupying half of the binding sites); its value is usually close to the value of K_m, the Michaelis constant of an enzymatic reaction.
n = the Hill coefficient, which describes the cooperativity of ligand binding in the following way:

- *Noncooperative binding* ($n = 1$): The affinity of the enzyme for a ligand molecule is not dependent on whether or not other ligand molecules are already bound. The Michaelis–Menten kinetics are a typical noncooperative binding kinetics, as one can find from Equation (5.2) that $n = 1$. Another example of noncooperation is the binding of myoglobin with oxygen, which also has $n = 1$.

- *Positively cooperative binding* ($n > 1$): Once one ligand molecule is bound to the enzyme, its affinity for other ligand molecules increases. The binding of hemoglobin with oxygen is positively cooperative, with a Hill coefficient $n = 2.8$. In mitogen-activated protein kinase (MAPK) signaling, the sigmoidal stimulus-response curve has a Hill coefficient estimated to be between 4 and 5.
- *Negatively cooperative binding* ($n < 1$): Once one ligand molecule is bound to the enzyme, its affinity for other ligand molecules decreases. This phenomenon is uncommon compared with positively cooperative binding. However, it was found that type I scavenger receptor class B (SR-BI) binds with HDL and LDL, with Hill coefficients $n = 0.85$ and $n = 0.65$ at 37°C, respectively.

Figure 5.1(b) illustrates Hill functions (Equation (5.3)) with different n values. One sees that the curve with $n = 1$ represents the hyperbolic Michaelis–Menten kinetics, which have been introduced. As n increases, the curve becomes more and more sigmoidal. Take $n = 10$ as an example. The initial increase of the stimulus $[X]$ has little response, i.e., the output v has no apparent increase even though $[X]$ is already high. But the response increases rapidly once $[X]$ reaches a threshold concentration of about $[X] = 1$, so much so that the response becomes full immediately after the threshold is exceeded. This switch-like behavior is called ultrasensitivity — the response is very sensitive to a small change around the threshold. This property is very important in molecular biology.

Figure 5.1(b) implies that positively cooperative binding confers ultrasensitivity. The more cooperative the binding, the more sensitive the response. Cooperative binding is a special case of allostery, the regulation of an enzyme or other protein by binding an effector molecule at the protein's allosteric site (that is, a site other than the protein's active site). Effectors that enhance the protein's activity are referred to as allosteric activators, whereas those that decrease the protein's activity are called allosteric inhibitors.

5.5.2 Zero-Order Ultrasensitivity

In previous sections, posttranslational modifications (PTMs) were introduced. One important PTM is covalent modification, which refers to any of a diverse group of processes in which the initially synthesized structures of biopolymers, especially enzymes, proenzymes, or structural macromolecules, are enzymically modified by the breakage of covalent bonds or the addition of new covalently linked groups. Some covalent modifications are reversible. That is, a biomolecule may switch between the active state and the inactive state, and the interconversion is rendered by a set of control enzymes, some of which promote activation and the others promote deactivation. A typical example of reversible covalent modification is the phosphorylation and dephosphorylation cycle (PdPC). The addition of a phosphate group to a biomolecule is catalyzed by protein kinases. The removal of a phosphate group from a biomolecule is catalyzed by protein phosphatases. Because both groups of enzymes may coexist, the final result (the level of phosphorylation) depends on the relative strength of the

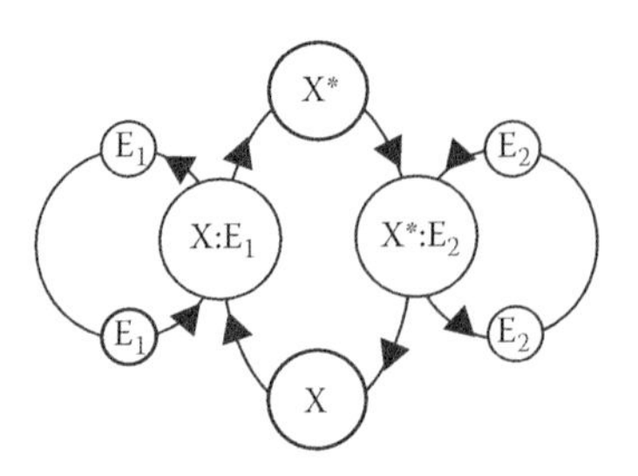

FIGURE 5.2 A schematic illustration of the covalent modification. Here X represents the biomolecule; X* represents the modified biomolecule; the enzyme E_1 promotes the modification; the enzyme E_2 reverses the modification.

two groups of enzymes. A covalent modification can be schematically represented by Figure 5.2.

Albert Goldbeter and Daniel E. Koshland studied the kinetics of covalent modification by using mathematical modeling [11]. The mathematical model was based on a pair of Michaelis–Menten kinetics, one for the modification and the other for the reverse modification:

$$E_1 + X \underset{d_1}{\overset{a_1}{\rightleftarrows}} E_1X \xrightarrow{k_1} E_1 + X^* \tag{5.4}$$

$$E_2 + X^* \underset{d_2}{\overset{a_2}{\rightleftarrows}} E_2X^* \xrightarrow{k_2} E_2 + X \tag{5.5}$$

where X represents the protein substrate; X^* represents the modified protein; E_1 and E_2 represent the converter enzymes. Because of the irreversibility, there is no apparent product (P) in the reaction, unlike the general enzyme kinetics Equation (5.2). For the case of PdPC, X and X^* represent the dephosphorylated and phosphorylated proteins, respectively; E_1 represents a group of protein kinases; and E_2 represents a group of phosphatases.

The dynamics of the system can be described by

$$\frac{d[X]}{dt} = -a_1[X][E_1] + d_1[E_1X] + k_2[E_2X^*], \tag{5.6}$$

$$\frac{d[E_1X]}{dt} = a_1[X][E_1] - (d_1 + k_1)[E_1X], \tag{5.7}$$

$$\frac{d[X^*]}{dt} = -a_2[X^*][E_2] + d_2[E_2X^*] + k_1[E_1X], \tag{5.8}$$

$$\frac{d[E_2X^*]}{dt} = a_2[X^*][E_2] - (d_2 + k_2)[E_2X^*], \tag{5.9}$$

$$X_T = [X] + [X^*] + [E_1X] + [E_2X^*], \tag{5.10}$$

$$E_{1T} = [E_1] + [E_1X], \tag{5.11}$$

$$E_{2T} = [E_2] + [E_2X^*], \tag{5.12}$$

where the brackets $[\cdot]$ signify concentrations. The symbols X_T, E_{1T}, and E_{2T} represent the concentrations of total substrate X, total enzyme E_1, and total enzyme E_2, respectively. Without loss of generality, we consider the case that X_T is in much excess over E_1 and E_2 so that the compound concentrations $[E_1X]$ and $[E_2X^*]$ can be neglected in Equation (5.10). That is, we assume $[E_1X] \approx 0, [E_2X^*] \approx 0$, which renders

$$X_T = [X] + [X^*].$$

In the steady states, the left-hand sides of Equations (5.6–5.9) are zero, and therefore Equations (5.6–5.12) reduce to algebraic equations. After simplification, one obtains a second-degree equation:

$$a\left(x^*\right)^2 + bx^* + c = 0, \tag{5.13}$$

where

$$\begin{aligned}
x^* &= [X^*]/X_T, \\
a &= \frac{V_1}{V_2} - 1, \\
b &= K_2\left(\frac{K_1}{K_2} + \frac{V_1}{V_2}\right) - \left(\frac{V_1}{V_2} - 1\right), \\
c &= -K_2\frac{V_1}{V_2},
\end{aligned}$$

and

$$\begin{aligned}
V_1 &= k_1E_{1T}, \\
V_2 &= k_2E_{2T}, \\
K_1 &= \frac{K_{m1}}{X_T} = \frac{d_1 + k_1}{a_1X_T}, \\
K_2 &= \frac{K_{m2}}{X_T} = \frac{d_2 + k_2}{a_2X_T}.
\end{aligned}$$

Note that the lower case x^* represents the mole fraction of the modified molecule; the lower case x represents the mole fraction of the unmodified protein. One must have

$$x + x^* = 1.$$

V_1 and V_2 are the rates of modification and reverse modification, respectively. V_1/V_2 thus quantifies the relative dominance of the two opposing enzymes. It will be used as the input stimulation to arouse response of the system. K_1 and K_2 are the Michaelis constants K_{m1} and K_{m2} divided by X_T, respectively. From Equation (5.13) one can actually solve x^* and x:

$$x^* = \frac{\sqrt{b^2 - 4ac} - b}{2a}, \tag{5.14}$$

$$x = \frac{2a + b - \sqrt{b^2 - 4ac}}{2a}. \tag{5.15}$$

Figure 5.3(a) gives a family of curves of x^* versus V_1/V_2. Figure 5.3(b) gives a family of curves of x versus V_1/V_2. For all the curves, the condition $K_1 = K_2 = K$

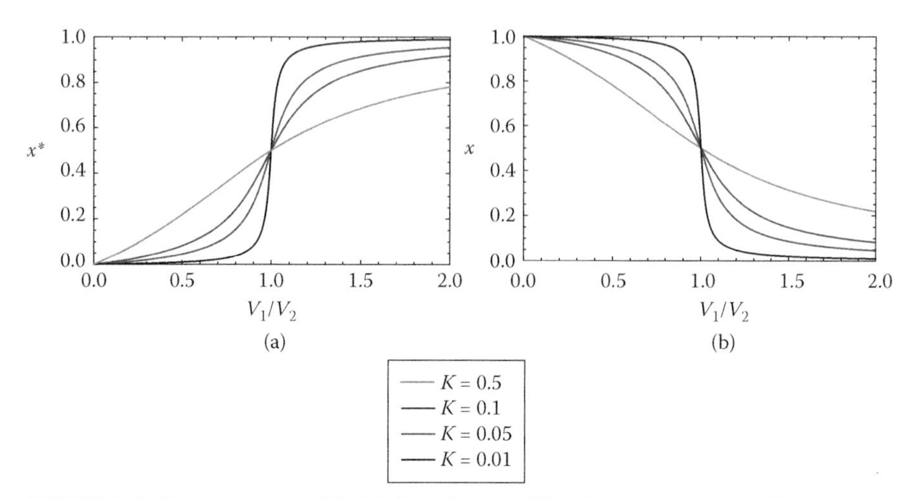

FIGURE 5.3 Fractions of modified (x^*) and unmodified (x) proteins at steady state as a function of V_1/V_2, the ratio of the modification rates. (a) The curves x^* versus V_1/V_2, which was obtained from Equation (5.14) with four values of K. (b) The curves x versus V_1/V_2, which was obtained from Equation (5.15) with four values of K. During the calculations, $K_1 = K_2 = K$ was assumed.

is assumed for the sake of simplicity. These curves have different sensitivity to the input V_1/V_2. Apparently, the sensitivity increases as K decreases. When $K = 0.5$, the curve represents a graded, slowly increasing response. When $K = 0.1$, the curve is more switch-like, but the sensitivity is still low. When $K = 0.01$, the curve becomes highly sensitive: for the range $V_1/V_2 < 1$, the fraction of modified protein x^* is nearly zero; for the range $V_1/V_2 > 1$, the fraction of modified protein x^* is nearly one; and a switching occurs abruptly around $V_1/V_2 = 1$.

Sensitivity rendered by the smallness of K exhibited in covalent modification is mechanistically different from sensitivity rendered by positive cooperativity exhibited in allosteric binding. Because

$$K = \frac{K_m}{X_T}, \tag{5.16}$$

the smallness of K can be realized by reducing K_m or by increasing X_T. The Michaelis constant K_m is relatively constant for a particular pair of interacting molecules. It may change with factors such as temperature. But the internal temperature of an organism is usually tightly regulated (37°C for humans). Therefore, the smallness of K actually corresponds to the largeness of X_T, the total concentration of the protein. The large amount of the protein would saturate the surfaces of the converter enzymes E_1 and E_2, for which reason the concept "enzyme saturation" has been developed to explain this ultrasensitivity.

To explain why this kind of ultrasensitivity is preceded by "zero-order," we consider a single Michaelis–Menten reaction, instead of the covalent modification that is a pair of Michaelis–Menten reactions. The kinetics have been presented in Equation (5.2). Figure 5.4 illustrates the reaction rate v as a function the substrate concentration $[X]$. If the total concentration X_T is small (the enzymes are thus not saturated), then

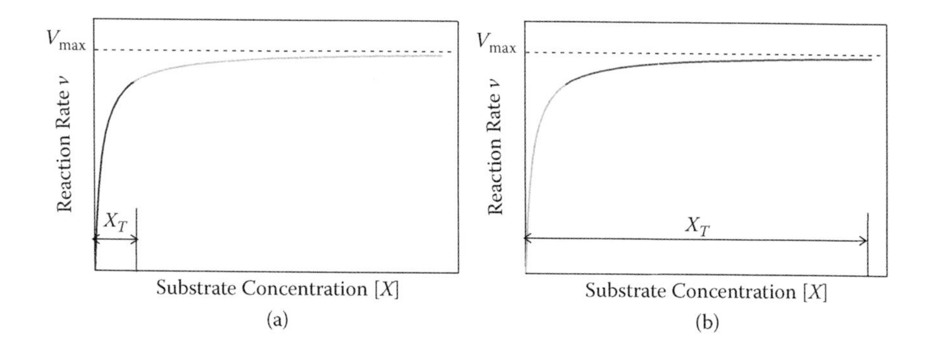

FIGURE 5.4 (a) The blue color indicates the first-order region of the substrate concentration $[X]$. (b) The red color indicates the zero-order region of the substrate concentration $[X]$.

the slope of $v([X])$ is relatively large (indicated by the blue color in Figure 5.4(a)); and we say that the kinetics are of first-order. If the total concentration X_T is large (enzyme saturation), then the slope of $v([X])$ is nearly zero for a wide range of $[X]$ values (indicated by the red color in Figure 5.4(b)), and we say that the corresponding kinetics are of zero-order. Now that enzyme saturation is characterized by zero-order kinetics, ultrasensitivity rendered by enzyme saturation (i.e., X_T is large) is known as zero-order ultrasensitivity.

We now consider covalent modification when both E_1 and E_2 are saturated. Under the condition $V_1/V_2 > 1$, the production rate of X^* is insensitive to the change of X for a long time (as long as X still saturates the enzyme E_1). Modification proceeds persistently in one direction without let-up until finally X cannot saturate E_1 any more. A steady state is reached at that point. Although the steady-state X may be not so small compared with E_1, the ratio $x = X/X_T$ is nearly zero and $x^* = X^*/X_T$ is nearly one (except in the cases $V_1/V_2 \approx 1$). Under the condition $V_1/V_2 < 1$, the steady state is characterized by $x \approx 1$ and $x^* \approx 0$ (except in the cases $V_1/V_2 \approx 1$). Under the condition $V_1/V_2 = 1$, one has $x = 0.5$ and $x^* = 0.5$. And a switching from nearly zero to nearly one occurs around $V_1/V_2 = 1$ (see, e.g., the black curve in Figure 5.3).

5.6 ZERO-ORDER ADAPTIVITY

Adaptation is the evolutionary process whereby an organism becomes better able to live in a new environment. Because the external environment (such as ambient temperature) and internal conditions (such as development, pregnancy, etc.) change frequently, the organism must possess a certain plasticity in adaptation. That is, the organism should be able to adjust its metabolic profile and physiologic traits to cope with various external or internal changes. Adaptation is certainly not a one-way process toward a fixed phenotype, but an evolving process responding to the changing environment.

Ultrasensitivity, as an important physiologic trait, also needs to be adjusted in response to the changing environment. Take nutrient uptake as an example. To survive stressful metabolic conditions such as overnutrition, starvation, pregnancy, and inflammation, an organism needs to adjust its sensitivity to nutrients. A starving cell needs to enhance its sensitivity so that it can get as much as possible. In an over-nutritional environment, a cell tends to reduce its sensitivity to avoid taking in too much.

Because ultrasensitivity can be achieved by both cooperative binding and enzyme saturation, it is natural to wonder which mechanism is easier to adjust ultrasensitivity. Here I argue that zero-order ultrasensitivity (by enzyme saturation) is much easier to adjust.

First, the degree of zero-order sensitivity is easy to tune. According to Equation (5.16), sensitivity can be easily tuned by changing X_T, the total concentration of the protein substrate. An increase of X_T would immediately reduce K and enhance sensitivity. A decrease of X_T would immediately increase K and reduce sensitivity. To change X_T, the cell can change the rate of mRNA expression of the protein substrate, which is a common mechanism of gene regulation. In contrast, sensitivity rendered by allosteric cooperation is largely fixed, because protein structure and conformation change are usually beyond regulation. This is demonstrated by the constant Hill coefficient of a specific binding process. For example, the binding of hemoglobin with oxygen has a Hill coefficient $n = 2.8$, which is difficult to alter. Enzyme saturation, through altering the substrate concentration, can finely tune the degree of sensitivity.

Second, the location of zero-order sensitivity is easy to tune. Here "location" refers to the switch point, namely, the concentration of the controlling enzyme at which the switching occurs. In Figure 5.3, the switch point is normalized, $V_1/V_2 = 1$. It is equivalent to

$$k_1 E_{1T} = k_2 E_{2T}. \tag{5.17}$$

Because E_{1T} and E_{2T} are enzyme concentrations, Equation (5.17) suggests that the switch-on point can be controlled by changing enzyme concentrations. Let E_{1T} be proportional to a control signal (input) I:

$$E_{1T} = \beta I. \tag{5.18}$$

By combining Equation (5.18) with Equation (5.17) and solving I, one obtains the switch point

$$I_{\text{on}} = \frac{k_2 E_{2T}}{k_1 \beta}.$$

Here one sees that the switch point is tunable, either by changing β or by changing the concentration E_{2T}. In Chapter 9, β and E_{2T} are related to the activities of PI3K and PP2A, respectively, which are indeed highly tunable. Zero-order sensitivity thus allows for a well-regulated switch point. In contrast, the switch point of cooperative binding is defined by the dissociation constant K in Equation (5.3), which is difficult to alter. Although K may change with temperature and pH, these parameters are fairly constant within the body. Therefore, the switch point is difficult to adjust through cooperative binding.

Adaptivity conferred by enzyme saturation is termed *zero-order adaptivity*. Because the condition of enzyme saturation can be adjusted by concentration changes and it is relatively easy to regulate concentrations, zero-order adaptivity appears to be easier to render than allosteric binding. In Chapter 9, we will find that the PI3K-AKT-TOR pathway uses these mechanisms to adjust its sensitivity to growth factors such as insulin. Or, at a more detailed level, X_T, β, and E_{2T} correspond to the concentrations of the total AKT, PI3K, and PP2A, respectively. The change of total AKT tunes the degree of sensitivity. The changes of PI3K and PP2A adjust the switch point.

6 Biomolecular Network

Biochemical reactions do not occur in isolation. A molecule (DNA, RNA, protein, etc.) in a living cell constantly interacts with a set of many other molecules, each of which has its own interacting partners. It is the complex network of interactions between all the molecules in the cell, as well as many external growth factors, hormones, cytokines, and chemokines, that collectively defines the cell's state and all of its biological functions. This giant network is, however, too large and too complex to analyze. It is not necessary to study the molecular interactions all together, because a specific biological function is primarily determined by only a subset of biomolecules in the cell. Therefore, a biomolecular network usually refers to the interactions among a subset of molecules, which collectively realize a specific biological function.

6.1 STRUCTURAL PROPERTIES

The molecular interactions can be graphically represented by a network, where nodes of the network represent biomolecules and edges of the network represent molecular interactions. There are generally two kinds of molecular interactions: stimulation and inhibition. An inhibitory (stimulatory) edge is usually represented by a red (green) arrow. Alternatively, inhibition can be represented by the edge $\dashv$ and stimulation can be represented by a usual edge $\rightarrow$. The network description allows for a ready transfer of concepts and techniques that have developed in fields such as mathematics (graph theory), physics, sociology, and engineering. For example, concepts such as modularity and tolerances are well known in engineering, and may be useful to analyze biological networks as well. Indeed, it is widely agreed that biomolecular networks share at least two important structural properties with engineering systems: modularity and use of recurring circuit elements [1]. In the following, we give a brief introduction to the two properties, as well as some other structural properties.

6.1.1 MODULARITY

A module consists of a set of components that work collectively to realize some function. A typical example of a module is a subroutine in software. It is not surprising that a biomolecular network has many modules. Indeed, a given network itself can be regarded as a module of the entire network of molecular interactions in the cell. As such, a given network may consist of several modules that each perform a specific biological function.

A module has defined output nodes that control the interactions with the other part of the network (the components complementary to the module). It also has defined input nodes that interact with each other but seldom interact with the outside nodes. This setting makes the module relatively separable from its surrounding structure. Modules convey an advantage in situations where the design specifications need to change often. In engineering, new devices or software can be easily constructed from existing, well-tested modules. A nonmodular device, in which every component is optimally linked to every other component, is effectively frozen and cannot evolve to meet new optimization conditions.

The modular design is especially advantageous for evolution. The changing environment exerts a selection pressure upon the cell, which makes the internal world of the cell more and more complex. New biological functions often need to be developed to cope with new challenges both within and without. These new biological functions are largely provided by old molecules (it would be expensive and impractical to develop a new set of molecules just to generate a new function). Therefore, it is more practical to rewire the network connection so that new functions can be developed. Modern genetic studies have shown that network rewiring has a high rate of survival and does confer evolvability [17]. For example, a link from molecule A to molecule B can be created by the promoter-A:ORF-B fusion, where ORF stands for open reading frame. The rewiring requirement is easier to fulfill in a modular setting: the module can be maintained largely intact, while the input and output nodes can be rewired or connected to new modules.

6.1.2 Recurring Motifs

The above discussion of modularity is actually about the temporal conservation of a module, namely, the relative constancy of the internal structure of the module, while the input/output nodes are subject to frequent rewiring during evolution. Then, how about the spatial conservation of a module? Is a given module used only in a specific network, or is it frequently used by other networks? Is it the unique design of a specific organism, or is it used also by other species?

It is natural to believe that the same modular design is repeatedly used, possibly with small modifications, in different biological networks and even in different organisms. This is easy to understand when one uses an analog in engineering. To construct a complex electronic circuit, for example, one does not have to make every connection from scratch. It would be more sensible to use existing electronic components (amplifiers, diodes, integrated circuits, etc.) to build a circuit efficiently. Similarly, a protein is often constructed with common patterns at different levels. For the secondary structure, the alpha helix and beta sheet are the dominant motifs. For the tertiary structure, a protein domain is a compact and stable three-dimensional structure that has been folded independently. Protein domains are conserved parts of a given protein that can evolve, function, and exist independently of the rest of the protein chain.

It is thus not strange to see that molecular interactions also display the same organizing principle, namely, the use of key wiring patterns repeatedly in a wide variety of biomolecular networks. Such a wiring pattern is often called a network motif, which is a network module that can perform a specific information processing task such as filtering out spurious input fluctuation, generating temporal programs of expression, or accelerating the throughput of the network. These network motifs were found in both prokaryotes and eukaryotes, and at both genetic transcription and protein signaling levels. The phosphorylation and dephosphorylation cycle can be regarded as a network motif that confers zero-order ultrasensitivity and zero-order adaptivity. In the following, three main types of regulatory motifs are briefly described: open loop, feedforward, and feedback. Because these regulatory motifs are ubiquitous control mechanisms beyond merely biology, the examples given below are not necessarily in the context of a biological network.

6.1.3 Regulatory Motifs

An *open-loop* controller computes its input into a system using only the current state and its model of the system. It does not use feedback to determine whether or not its output has achieved the desired goal. As a consequence, an open-loop system cannot correct any errors that it could make. Open-loop control is usually used to realize tasks in which high precision is not required.

Feedforward. A feedforward system can be characterized by two regulatory nodes (nodes 1 and 2) and one target node (node 3), with node 1 regulating node 2 and both nodes 1 and 2 jointly regulating node 3. For example, nodes 1 and 2 may represent two transcription factors and node 3 may represent the target gene. Because of the two apparent kinds of molecular interaction (stimulation and inhibition), the feedforward loops in molecular biology have $2^3 = 8$ types (Figure 6.1). It was found that four types (termed incoherent feedforward loops) act as sign-sensitive accelerators and the other four types (termed coherent feedforward loops) act as sign-sensitive delays [21]. The incoherent feedforward loops speed up the response time of the target gene expression following stimulus steps in one direction (e.g., off to on) but not in the other direction (on to off). Likewise, the coherent feedforward loops delay responses to stimulus steps, but only in one direction.

Feedback, as part of a chain of cause and effect that forms a circuit or loop, is a process in which information about the past or the present influences the same phenomenon in the present or future. Feedback is commonly divided into two types — positive feedback and negative feedback.

Negative feedback control is the most important and ubiquitous regulatory mechanism in nature. In a negative feedback system, the result of a process influences the operation of the process itself in such a way as to reduce changes. Negative feedback

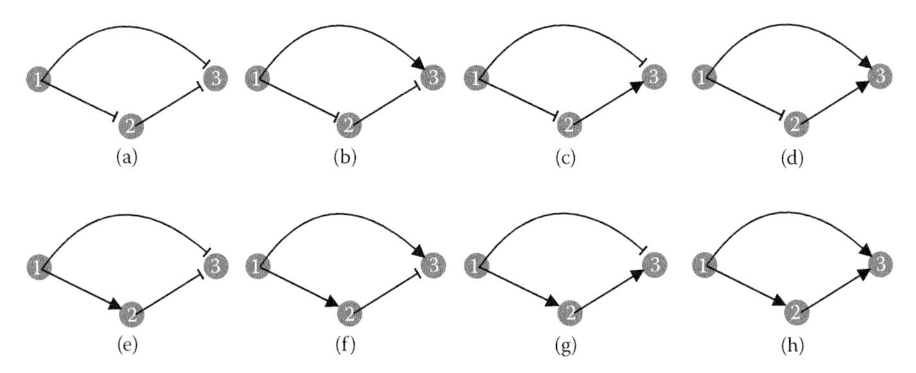

FIGURE 6.1 Feedforward loops can be represented by the interactions between three nodes 1, 2, and 3. Here node 2 is regulated by node 1 and node 3 is regulated by both nodes 1 and 2. There are eight types of feedforward loops. Inhibition is represented by $\dashv$ and stimulation is represented by $\rightarrow$.

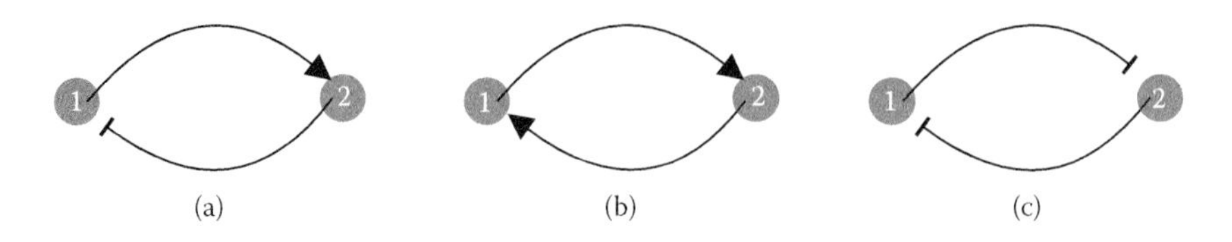

FIGURE 6.2 Feedback loops formed by nodes 1 and 2. (a) Negative feedback loop. (b) Positive feedback loop (mutual stimulation). (c) Double negative feedback loop (mutual inhibition).

tends to make a system self-regulating; it can produce stability and reduce the effect of fluctuations. Negative feedback is found in almost all scientific disciplines and is widely used in engineering. It occurs naturally within living organisms. The glucose-insulin system is a typical negative feedback one: plasma glucose stimulates the production of insulin, which in turn lowers the glucose level. In molecular biology, negative feedback can be represented by two molecules (nodes 1 and 2); node 1 stimulates node 2 which in turn inhibits node 1 (Figure 6.2(a)).

Positive feedback is a process in which the effects of a small disturbance on a system induce an increase in the magnitude of the perturbation. A positive feedback can be represented by two nodes 1 and 2: node 1 stimulates node 2 which in turn stimulates node 1 (Figure 6.2(b)). Alternatively, node 1 inhibits node 2 which in turn inhibits node 1 (Figure 6.2(c)). The latter is known as double negative feedback, which is equivalent to positive feedback. Positive feedback is well studied in gene regulation and cell signaling networks, where it is most often associated with bistability. It has been proved that positive feedback is a necessary condition for generating bistability [7]. Moreover, a genetic toggle switch was realized through the construction of a double negative feedback loop [16], which has become one of the best-known examples in synthetic biology. In Chapter 3, we have shown that bistability is necessary for the optimal control of glucose homeostasis. The optimal control is realized by a molecular circuit known as the PI3K-AKT-TOR pathway, which is described in detail in Chapter 7. In the circuit there does exist a positive feedback loop that is essential in producing hysteresis and bistability.

Feedback loops exist ubiquitously in biological systems, including biomolecular networks. Figure 6.3 shows a simplified network of the budding yeast cell cycle, which has 11 nodes and 34 edges. As usual, inhibitory edges are represented by $\dashv$ and stimulatory edges are represented by $\rightarrow$. Ten of the 34 edges form into various feedback loops: negative feedback (colored in blue), positive feedback (colored in green), and double negative feedback (colored in red).

Figure 6.3 is a good example to demonstrate the structural properties of biological networks. First, it demonstrates that nature likes to use recurring motifs, because about one-third of the edges in the network form into feedback loops. Second, it demonstrates modular design. After removing the feedback loops, the remaining 24 edges (colored in gray) can still carry out the basic cell cycle steps [36]. In fact, these edges constitute a minimal network — the smallest network capable of providing the basic cell cycle function. Therefore, the network can be decomposed into a backbone module that realizes the basic steps of cell cycle and several feedback motifs to regulate the cell cycle.

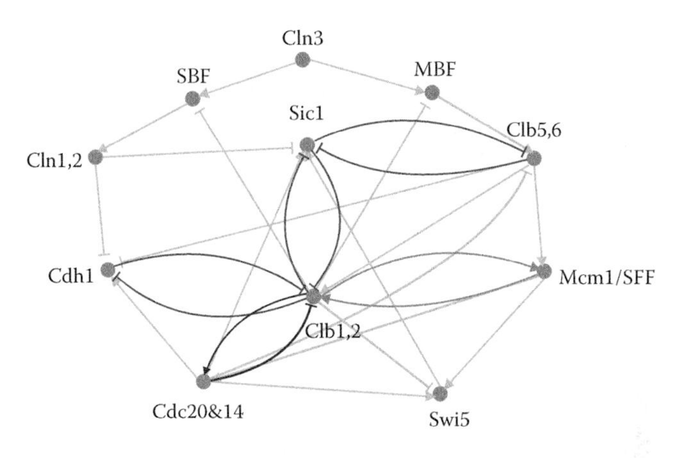

FIGURE 6.3 Feedback loops in the budding yeast cell cycle network. The negative feedback, positive feedback, and double negative feedback loops are colored in blue, green, and red, respectively. See color insert.

6.1.4 Scale-Free Property

A scale-free network is characterized by a (asymptotic) power law degree distribution, where the degree of a node is the number of connections the node has to other nodes, and the degree distribution is the probability distribution of these degrees over the whole network. The power law distribution implies that the fraction $P(k)$ of nodes in a network having k connections has the following expression when k is large:

$$P(k) \sim ck^{-\gamma},$$

where c is a normalization constant and γ is a parameter whose value is typically in the range $2 < \gamma < 3$.

Many biological networks were found to be scale-free and to have $2 < \gamma < 3$. Other real-world networks that are thought to be scale-free include social networks (e.g., the collaboration of movie actors in films), financial networks (e.g., interbank payment networks), semantic networks, airline networks, and the World Wide Web.

The most notable characteristic of scale-free networks is the presence of "hubs," namely, those nodes with a degree that greatly exceeds the average. The hubs are thought to serve specific purposes in their networks. The clustering coefficient distribution is another important characteristic of scale-free networks. It decreases as the node degree increases, and the distribution also follows a power law. This implies that the low-degree nodes belong to very dense subnetworks and those subnetworks are interconnected through hubs. Finally, the average distance between two nodes also characterizes a network. It turns out that the distance is small for a scale-free network. Therefore, a scale-free network is usually also a small-world one. Notably, a subnetwork-containing, scale-free network has an ultrasmall diameter $d \sim \ln \ln N$, where N is the number of nodes in the network. The diameter of a growing scale-free network might be considered almost constant in practice.

6.2 DYNAMICAL PROPERTIES

Biomolecular networks are dynamical. A cell is a dynamical unit, consisting of a large number of molecules whose combined actions determine the physiology of the cell. The dynamical properties of a cell are implicit in the network of interactions among the individual molecules. This giant network determines how a cell changes in space and time to respond to stimuli; how a cell grows, differentiates, divides; how a cell makes the decision to die or not.

Biomolecular networks are nonlinear. Nonlinearity exists naturally even in the simplest enzymatic reaction: the initial change of the reaction rate is approximately linear but the reaction rate levels off finally as the substrate concentration becomes large (see Figure 5.4). This enzymatic reaction can be described by the Michaelis–Menten kinetics with a Hill coefficient $n = 1$ (Equation (5.2)). By including mechanisms such as cooperative binding and enzyme saturation, the reaction becomes more complex and nonlinearity becomes greater, as demonstrated by an increased Hill coefficient and the onset of extraordinary properties such as ultrasensitivity. Nonlinearity can be further augmented by the complex network of molecular interactions. Even a positive feedback loop, one of the simplest network motifs, can generate remarkable properties such as hysteresis and memory.

6.2.1 DYNAMICAL CHANGE AND ATTRACTOR

Biomolecular networks can be modeled by high-dimensional dynamical systems, usually in the form of coupled ordinary differential equations. The quantities of a dynamical system can be divided into two categories.

The first category includes all the parameters in the system. These parameters are usually constant, to model entities that are relatively invariant (such as rate constants). The second category includes all the variables in the system. The variables are usually the concentrations of the node molecules of the biomolecular network under consideration. These variables collectively define the network state and are thus known as the state variables.

At a specific time t, both the parameters and the values of the state variables determine how fast each molecule changes with time, namely, how much each concentration will change in the next small increment of time Δt. The network state at time $t + \Delta t$ can thus be determined. Therefore, the entire evolution of the network state can be determined if the initial values (at time $t = 0$) of the state variables are given. The determination is usually realized by numerically integrating the differential equations.

But what does the dynamical change tend toward? According to the mathematical theory of dynamical systems, the network state has four destinies, or four attractors in mathematical terms. In the following, the four types of attractor are briefly described in terms of a general dynamical system that is not limited to biological networks.

- *Steady state* (fixed point) is the final state of dynamical evolution, at which every component in the system does not change in concentration. For example, the glucose-insulin feedback system is a typical dynamical system at the organismal level. In the fasting state, the plasma glucose and insulin concentrations are quite stable and can thus be regarded as steady states. It should be noted that the constancy of the steady states is in terms of a given set of parameter values. If the parameters change, the steady states change accordingly.

- *Limit cycle* is a periodic orbit of the system that is isolated (i.e., attracting nearby orbits). Limit cycles imply self-sustained oscillations. Any small perturbation from the limit cycle would cause a return of the system state to the trajectory, making the system state stick to the limit cycle. The periodic motion of an ideal pendulum, however, is not a limit cycle attractor, because its orbits are not isolated. In the phase space of the ideal pendulum, near any point of a periodic orbit there is another point that belongs to a different periodic orbit. The orbit is thus not a limit cycle, because it does not attract the latter orbit.
- *Limit torus*. As a periodic trajectory, a limit cycle may have multiple frequencies of oscillation. These frequencies must be commensurate with each other, in the sense that any two frequencies must form a rational fraction. If there exists a pair of frequencies that form an irrational fraction, the trajectory is no longer closed, and the limit cycle becomes a limit torus.
- *Strange attractor* has a noninteger dimension. This is often the case when the underlying dynamics are chaotic. But there also exist strange attractors that are not chaotic. Strange attractors are often differentiable in a few directions. Some strange attractors, such as Cantor dust, are not differentiable.

The existence of attractors depends on the modeling approaches used. If a discrete method is used to model a deterministic system, then the number of states is finite, and limit tori and strange attractors cannot possibly exist as attractors. Because the states are finite in number, the deterministic system either stays at a particular state forever (that state is thus a steady state) or leaves a state for a while but will finally return (that state is only one point along a cycle). A deterministic Boolean network model, for example, has only 2^N network states, where N is the number of nodes in the network. Depending on the network topology and dynamical rules, the network state either converges to a steady state or goes through a limit cycle repeatedly.

Steady states are the dominant attractors in biology. Due to the homeostasis principle (internal constancy), biological systems usually function at the steady state. Limit cycles are less common, but do exist in biology. For example, heartbeat and cell cycle dynamics can be modeled as limit cycles. Limit tori and strange attractors rarely exist in biology. For this reason, the limitation of the Boolean modeling approach (in the sense that it does not produce limit tori and strange attractors) is not biologically relevant.

6.2.2 Stability

The stability of an attractor involves the behavior of the state of a dynamical system. Let the system state be at the attractor initially. With a small, transient perturbation, the system state deviates from the attractor. If the system state leaves the attractor forever, no matter how small the perturbation is, then we say that the attractor is unstable. If the system state returns to the attractor if the perturbation is sufficiently small, then we say that the attractor is stable. If an attractor is unstable, then it does not exist in practice, because in the real world noisy perturbations always exist. For a stable attractor, stability can be measured by quantities such as the Lyapunov exponent, which is the rate of separation of very close trajectories (by definition, infinitesimally close). The larger the absolute value of the Lyapunov exponent, the more stable the system is. Stability is an important dynamical property in biology, because biological systems function against the backdrop of extensive molecular fluctuations. A common method to confer stability is the use of negative feedback loops, which buffer against random perturbations caused by fluctuations.

6.2.3 MONOSTABILITY AND MULTISTABILITY

Monostability refers to the usual phenomenon that a set of system parameters only corresponds to one steady state. When the set of parameters changes, the steady state also changes, but the number of steady states is still one. Multistability refers to the nonlinear phenomenon of existing multiple steady states for the *same* set of system parameters.

Figure 6.4 shows the change of the state variable x with the parameter λ. The monostable portion is in the ranges $[0, \lambda_1)$ and (λ_2, ∞). The continuous increase of λ makes x slide along a continuum of steady states, and a given λ value corresponds to only one steady state. The multistable portion is in the range $[\lambda_1, \lambda_2]$. In that range, a single λ corresponds to three steady states: two stable (solid lines) and one unstable (the dashed line). The unstable state does not exist in reality, because a slight perturbation would destroy the steady state. Therefore, this example actually represents bistability, essentially the same bistability as the one introduced in Part I. Strictly speaking, a multistable response is not a mathematical function. To qualify

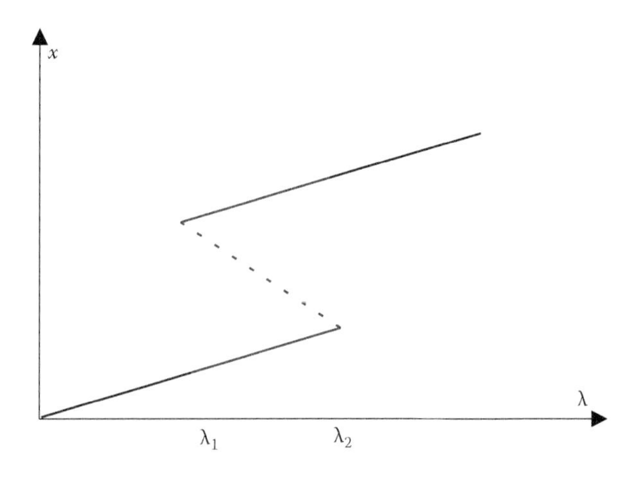

FIGURE 6.4 The steady state x as a "function" of λ. The multistability region is between λ_1 and λ_2 and the remaining is the monostability region.

as a function, one preimage (λ here) must always correspond to only one image (x here). But in the multistable region, one preimage corresponds to three images.

6.2.4 Sensitivity

Sensitivity may refer to several related but different concepts. In electronics, for example, sensitivity refers to the minimum magnitude of input signal required to produce a specified output signal having a specified signal-to-noise ratio or other specified criteria. It is a quality parameter of electronic devices such as microphones, loudspeakers, and hydrophones.

Even in biology, the meaning of sensitivity may be quite different. In enzyme kinetics, sensitivity (or ultrasensitivity) refers to the switch-like behavior, namely, the sigmoidally shaped response (unresponsive to a subthreshold stimulus but fully responsive once the threshold is exceeded). However, sensitivity is often used in biology as a synonym for responsiveness. For example, "insulin sensitivity" is a well-known term that actually means the responsiveness of tissue cells to the insulin stimulus. It does not refer to the switch-like response of a molecule to the insulin concentration, even though the response does behave in that manner (and that is what causes confusion). As such, insulin hypersensitivity does not refer to an overly sharp switch behavior but refers to the phenomenon that an extremely small insulin concentration can induce a dramatic response.

To avoid confusion, in this book we avoid the use of "insulin sensitivity" as much as possible and prefer to use "insulin responsiveness" instead. The word sensitivity is used exclusively for the switch-like behavior, conferred by mechanisms such as cooperative binding and enzyme saturation.

6.3 CENTRAL CHALLENGES

The ultimate goal of molecular biology is to understand the intricate relationship among network structure, dynamical properties, and biological function. This problem becomes especially important in the era of systems biology because a biomolecular network usually consists of a large number of interacting molecules.

The majority of works in systems biology are bottom-up (network-based). They start from a particular network to deduce higher-level properties (network structure → dynamical property → biological function). For example, the identification of the scale-free structural property and research thereafter have revealed the dynamical and functional consequences of the scale-free property. It turns out that the scale-free property strongly correlates with a network's robustness to failure. In a scale-free network, the major hubs are closely followed by smaller ones. These smaller hubs, in turn, are followed by other nodes with an even smaller degree, and so on. This hierarchy allows for a fault tolerance capacity. If failures occur at random and the vast majority of nodes are those with small degree, the likelihood that a hub would be affected is almost negligible. Even if a hub failure occurs, the network will generally not lose its connectedness, due to the remaining hubs. The studies of feedback motifs also lead to the elucidation of their dynamics and functions. It is well known that negative feedback regulation can either lead to steady state and thus homeostasis, or lead to periodic motion so that certain biological functions can be realized. On the other hand, positive feedback is known to produce hysteresis (sustained responses from transient stimuli). Such a response can render cellular memory — the stimulus vanishes but its effects last long, which is important for biological processes such as cell differentiation and cell cycle progression. Part III of this book primarily follows the bottom-up approach. Based on the mathematical analysis of the PI3K-AKT-TOR network, the range of dynamical behaviors of the network will be revealed and then mapped to physiologic and pathologic phenotypes.

The bottom-up approach has certain limitations despite its apparent importance. Due to the immense complexity of a molecular network, one may easily become overwhelmed by the blizzard of molecular details and may thus miss the forest for the trees. Great simplifications of the network have to be made in the first place. But it is not always apparent which parts of the network can be simplified and which parts cannot. Second, it is usually difficult to identify the necessary conditions at the bottom to realize a fine property at an upper level. For example, bistability proves to be an optimal control mechanism at the organismal (upper) level. To determine which structural feature of the molecular network (bottom) is responsible for generating bistability, one has to interrogate all the modules, motifs, and mechanisms in the network. The enumeration is time consuming and may well be incomplete. Even though positive feedback is found to generate bistability, it is difficult to rule out the possibility that other motifs and mechanisms can also do that. That is, it is difficult to prove the uniqueness of positive feedback in realizing bistability.

The limitations of the bottom-up approach highlight the importance of the top-down direction, which starts from biological functions to deduce how network dynamics and structure are constrained (function $\rightarrow$ dynamics $\rightarrow$ structure). The top-down approach allows one to derive the necessary properties of the molecular network from the constraints of the physiological system. Although this approach has been explored relatively less often, there are some impressive examples. For example, it was proved that positive feedback is a necessary condition for multistability (multistability $\rightarrow$ positive feedback) [7]. Therefore, positive feedback is indeed unique, as far as producing bistability is concerned. Part I of this book followed the top-down approach. Both optimality and homeostasis are overarching principles in physiology (biological function). Starting from the assumption of optimal homeostasis, we demonstrated that bistability is a necessary condition to confer optimal homeostasis. Together with the fact that positive feedback is necessary for bistability, a chain of top-down necessary conditions was established: optimal homeostasis (biological function) $\rightarrow$ bistability (dynamical property) $\rightarrow$ positive feedback (structural property).

7 PI3K-AKT-TOR Pathway

In Part I of this book, the biological processes of food intake and energy metabolism were introduced, as well as the homeostasis of glucose and insulin in the blood. We also learned that glucose enters a cell via a variety of glucose transporters (GLUTs). In particular, GLUT4 is responsible for massive glucose uptake by striated muscles and adipose tissues and the translocation of GLUT4 to the plasma membranes of these cells is mediated by insulin.

This chapter describes the detailed molecular events triggered by insulin stimulus, which ultimately render glucose transportation into the cell. Insulin, a peptide hormone with a molecular weight of 5808 Da, first binds with insulin receptors on the surface of a cell. The signaling is processed by a network of interactions among many molecules, collectively known as the PI3K-AKT-TOR pathway. The pathway has several key regulators, including PI3K, AKT, and TOR, which are described in the following sections. It should be stressed that the pathway actually functions as a whole and it is difficult to introduce one molecule without mentioning the other two. Therefore, each of the three sections actually describes one aspect of the pathway functioning. In the last section, a backbone structure is extracted from the pathway, as a summary of this chapter.

Finally, it is worth noting that glucose transportation is only one function of the PI3K-AKT-TOR pathway. In fact, the pathway plays many important roles, including cell growth, proliferation, and survival; cell migration and polarity; cell cycle progression; angiogenesis; and cell metabolism (lipid and glucose). As such, insulin is only one of the hormones that can stimulate the pathway. The pathway can also respond to insulin-like growth factors, epidermal growth factors, and many more.

7.1 PHOSPHOINOSITIDE 3-KINASE (PI3K)

The phosphoinositide 3-kinases (PI3Ks) belong to a conserved family of intracellular lipid kinases that phosphorylate the 3′-hydroxyl group of phosphatidylinositol and phosphoinositides. This phosphorylation leads to the activation of many intracellular signaling pathways that regulate cell metabolism, survival and polarity, and vesicle trafficking. The ancient role for PI3Ks is probably to mark specific cellular membranes for trafficking events. In yeast (a unicellular organism), only one form of PI3K has been found, vacuolar protein-sorting defective 34 (Vps34). The primary function of Vps34 is vesicle trafficking. In multicellular organisms, additional isoforms of PI3K have evolved for the dedicated purpose of signal transduction, including insulin-mediated metabolism. Over the past years, it has become evident that PI3Ks and their suppressor PTEN belong to the most highly mutated molecules in human cancers, underscoring their central role in human carcinogenesis.

PI3Ks can be grouped into three classes according to their substrate preference and sequence homology, each playing distinct roles in cellular signal transduction.

7.1.1 Class I PI3Ks

Class I PI3Ks are divided into two subfamilies, depending on the receptors to which they couple. Class IA PI3Ks are activated by growth factor receptor tyrosine kinases (RTKs), which include insulin receptors. Class IB PI3Ks are activated by G-protein-coupled receptors (GPCRs).

Class IA PI3Ks are heterodimers of a p85 regulatory subunit and a p110 catalytic subunit. In mammals, there are numerous isoforms of both p85 (p85α, p85β, p85γ, ...) and p110 (p110α, p110β, p110γ, ...). The p85 regulatory isoforms have a common core structure consisting of a p110-binding domain flanked by two Src-homology 2 (SH2) domains. The p110β isoform is regulated not only by the p85 regulatory subunit but also by binding to heterotrimeric G proteins. Therefore, it might integrate signals from GPCRs and RTKs. Although p85 is indispensable for PI3K signaling, mice lacking individual isoforms of p85 showed enhanced PI3K signaling. The reason appears to be that p85 is in excess of p110 in many cell types and that free p85 can form a cytoplasmic sequestration complex with insulin receptor substrate (IRS), thereby limiting the ability of IRS to activate PI3K at the membrane.

Class IA PI3Ks evolved early, during the transition from unicellularity to multicellularity. The social amoeba *Dictyostelium discoideum* has evolved a family of class IA PI3Ks. In *Caenorhabditis elegans*, the class IA PI3K regulates dauer formation, metabolism and longevity, by acting downstream of an insulin-like hormone. In *Drosophila melanogaster*, the class IA PI3K controls cell growth and proliferation. In mammals, class IA PI3Ks regulate glucose homeostasis, cell migration, growth, and proliferation.

Class IB PI3Ks are heterodimers consisting of a p101 regulatory subunit and a p110γ catalytic subunit. They appear to be exclusively activated by GPCRs through interacting directly with the G$\beta\gamma$ subunit of trimeric G proteins. They do not have p85 regulatory subunits and therefore are not regulated by RTKs. Therefore, this subclass is less relevant to insulin signaling.

7.1.2 Class II PI3Ks

Class II PI3Ks consist of only a single p110-like catalytic subunit. In vitro, Class II PI3Ks preferentially phosphorylate phosphatidylinositol and, to a lesser extent, phosphatidylinositol-4-phosphate (PI-4-P). However, they usually do not phosphorylate phosphatidylinositol-4,5-bisphosphate (PI-4,5-P2). Class II PI3Ks regulate membrane trafficking and receptor internalization, as indicated by their binding with clathrin and their location to coated pits. On the other hand, class II PI3Ks can be activated by RTKs, cytokine receptors, and integrins. However, their specific functions in response to these activators are not well understood. Class II PI3Ks appear in multicellular organisms that originated as early as *Caenorhabditis elegans*.

7.1.3 Class III PI3K

Class III PI3K consists of a single member, Vps34. Although originally identified in budding yeast as the gene product required for trafficking vesicles from the Golgi apparatus to the vacuole, Vps34 is present in every eukaryotic organism that has been examined thus far, and has a highly conserved function in regulating vesicle trafficking. Vps34 was recently found to regulate the mammalian target of rapamycin (mTOR) activity in response to amino acid availability. Therefore, this class might also be crucial for controlling cell growth and proliferation. In addition, Vps34 has been implicated in autophagy, which is a cellular response to nutrient starvation.

7.1.4 PI3K Signaling and Insulin Resistance

Class IA PI3K signaling has a certain tissue specificity. Defective PI3K signaling in the muscle results in impaired glucose uptake, hyperlipidemia, and adiposity. Defective PI3K signaling in the liver results in enhanced gluconeogenesis, hyperglycemia, hyperinsulinemia, and reduced circulating lipids.

Insulin resistance can be caused by attenuated PI3K signaling. Increased levels of p85 might contribute to muscle insulin resistance and diabetes. It has been demonstrated that human placental growth factor can cause insulin resistance by increasing the expression of the p85α monomer, which competes in a dominant negative fashion with the p85-p110 heterodimer for binding to the IRS-1 protein. This insulin resistance, if mild, is a beneficial physiological response and is necessary to divert fuels to the fetus to meet fetal growth demands. Elevated levels of p85 were also observed in muscles of individuals with type 2 diabetes. On the other hand, PI3K signaling can also be attenuated by serine phosphorylation of IRS proteins by numerous kinases, such as the c-Jun N-terminal kinase (JNK). JNK-mediated serine phosphorylation inhibits the tyrosine phosphorylation of IRS, thereby uncoupling IRS from PI3K. Obesity often leads to cellular oxidative stress, which enhances serine phosphorylation of IRS.

7.1.5 PI3K Signaling and Cell Growth

Organisms increase their size and cell number through the synthesis of new cellular components, including proteins, nucleic acids, and lipids. In metazoans (animals), cell growth is controlled by growth factors such as insulin-like growth factor 1 (IGF1). We and other vertebrates have distinct (although highly homologous) receptors for insulin and IGF1, and can therefore independently control acute glucose homeostasis and long-term cell growth and development. As a comparison, *Caenorhabditis elegans* and *Drosophila melanogaster* use the same receptor (a homologue of insulin and IGF1 receptors) to regulate both metabolism and development. Despite the difference at the first step of signaling, the downstream signaling components (including IRS and PI3K) are largely the same. Besides IGF1, other growth factor receptors (e.g., the platelet-derived growth factor receptor and the epidermal growth factor receptor) also activate PI3K signaling, which in turn promotes cell and tissue growth.

7.1.6 PI3K Signaling and Cancer

PI3K plays an important role in cancer, which is characterized by uncontrolled cell growth and proliferation. In fact, the very discovery of PI3K in the 1980s was due to its association with oncoproteins. About a decade later, the p110α catalytic subunit of PI3K was identified as an avian retrovirus-encoded oncogene that could transform chick embryo fibroblasts in vitro. *PIK3CA*, the gene that encodes p110α, is frequently amplified in many human cancers. Most point mutations in p110α cluster around two hotspots: E545 in the helical phosphatidylinositol kinase homology domain and H1047 near the end of the catalytic domain. The affected residues within the hotspots are highly conserved evolutionarily. *PIK3CA* mutations generally arise late in tumorigenesis, just before or coincident with invasion. Mutations in other p110 isoforms have not been identified in cancers, highlighting the oncogenic potential of p110α.

PI3K activity is suppressed by a tumor suppressor PTEN, which functions as a PIP3 3′-phosphatase. The gene *PTEN* was frequently found to be deleted in advanced cancers. The loss of *PTEN* may cause overactivity of PI3K and thus tumorigenesis. Indeed, the loss of *PTEN* occurs at a rate second only to *TP*53 in malignant tumors. Germline mutations in the *PTEN* gene often result in benign tumors with a high risk of cancer. Somatic (i.e., tumor-specific) mutations, on the other hand, occur in many glioblastomas, prostate cancers, breast cancers, and melanomas. The frequency of loss of heterozygosity for *PTEN* far exceeds that of biallelic inactivation, which indicates that haploinsufficiency for *PTEN* might suffice in promoting tumorigenesis in certain cellular contexts. This feature distinguishes *PTEN* from *TP*53, for which biallelic inactivation is necessary for tumorigenesis.

7.2 PROTEIN KINASE B (AKT)

The serine/threonine protein kinase B (PKB) is a master regulator of many important biological functions and processes, including cell growth and proliferation, cell survival, cell migration, glucose and lipid metabolism, skeletal muscle and cardiomyocyte contractility, angiogenesis, and stem cell self-renewal. PKB was originally identified as the human homologue of the viral oncogene v-akt. Therefore, the kinase was originally known as AKT and this term is still used more frequently than PKB. The "AK" in AKT was a temporary classification name for a mouse strain originally bred and maintained by Jacob Furth. The "T" in AKT stands for thymoma that was spontaneously developed in the mouse strain. Later, AKT was found to bear significant homology to both protein kinase A (PKA) and protein kinase C (PKC). AKT contains an N-terminal pleckstrin homology domain, a central kinase domain, and a C-terminal regulatory domain that contains the hydrophobic motif.

In mammalian cells, AKT has three closely related and highly conserved (greater than 80% sequence identity) homologues: AKT1, AKT2, and AKT3 (also known as PKBα, PKBβ, and PKBγ, respectively). AKT1 is more involved with cell growth and survival. AKT2 is more involved with metabolism. The role of AKT3 is less clear, though it appears to be predominantly expressed in the brain.

7.2.1 AKT PHOSPHORYLATION (ACTIVATION)

AKT is usually activated by growth factors (insulin, IGF1, etc.) through the PI3K signaling described above. Upon activation, PI3K generates phosphatidylinositol-

3,4,5-trisphosphate (PIP3) from phosphatidylinositol-4,5-bisphosphate (PIP2). PIP3 is a lipid second messenger that activates many downstream molecules by binding to their pleckstrin homology domains, including AKT.

Binding of PIP3 therefore leads to the membrane recruitment of AKT, which allows AKT to be phosphorylated by the enzyme PDK1 at threonine 308 (T308) in the central kinase domain. PDK1 is a 3′-phosphoinositide-dependent protein kinase. Its C-terminal pleckstrin homology domain binds phospholipids, keeping it constitutively localized at the plasma membrane. The interaction of the AKT pleckstrin homology domain with 3′-phosphoinositides is thought to impose conformational changes in AKT that allow the exposure of T308 and serine 473 (S473), which is within the hydrophobic motif of the C-terminal regulatory domain.

The full activation of AKT depends on the phosphorylation at S473, by TORC2 (TOR complex 2, which is TOR outfitting in the rictor complex). The direct homodimerization of the two pleckstrin homology domains between AKT and PDK1 might mediate protein proximity and subsequently phosphorylate T308 in AKT, which stabilizes the activation loop in an active conformation and renders S473 phosphorylation by TORC2.

Phosphorylations at T308 and S473 are both necessary and sufficient for the full activation of AKT. Point mutants at these two sites showed little AKT activity, even after stimulation with insulin or IGF1. On the other hand, the phosphorylation-mimicking mutant (T308D/S473D) showed constitutive AKT activation.

7.2.2 AKT Dephosphorylation (Deactivation)

The phosphate PP2A is the main enzyme to dephosphorylate the activated AKT. It consists of a catalytic subunit (PP2A/Cα or β), a regulatory or structural A subunit (PP2A/Aα or β), and a regulatory B subunit (PP2A/B, which has 16 isoforms) that determines substrate specificity. PP2A/A is composed of 15 nonidentical tandem repeats of a 39 amino acid sequence. Repeats 1–10 bind with PP2A/B and repeats 11–15 bind with PP2A/C. On the other hand, PP2A/C forms a complex with an array of PP2A/B that modulates its catalytic activity, substrate specificity, and subcellular localization. Because PP2A/B has 16 isoforms, it is believed that PP2A exercises regulatory flexibility and differential substrate specificity through the preferential association of the core dimer (PP2A/A and PP2A/C) with one of the regulatory PP2A/B. Indeed, PP2A/B is expressed differentially by tissue and temporally during development; therefore, it is the B subunit of PP2A that specifies substrate specificity and defines the physiological roles of PP2A. PP2A preferentially dephosphorylates AKT on the T308 site, although it can also dephosphorylate AKT on S473 under certain conditions.

PHLPP specifically dephosphorylates AKT on S473 but not T308. The PHLPP family comprises three members, PHLPP1α, PHLPP1β, and PHLPP2. The PHLPP proteins contain an identical domain structure, with a pleckstrin homology domain followed by a region of leucine-rich repeats, a PP2C phosphatase domain, and a C-terminal PDZ ligand. Both PHLPP1β and PHLPP2 also contain a Ras-associated domain that precedes the pleckstrin homology domain. Both the PP2C domain and the C-terminal PDZ-binding domain are required for PHLPP-mediated dephosphorylation of AKT on S473, while the pleckstrin homology domain of PHLPPs is dispensable. The AKT dephosphorylation neither destabilizes PHLPP nor promotes its degradation.

PHLPP-mediated AKT dephosphorylation is specific from both sides. It was found that PHLPP1 co-immunoprecipitates with only AKT2 and AKT3, whereas PHLPP2 co-immunoprecipitates with only AKT1 and AKT3. This implies that only PHLPP1 dephosphorylates AKT2, only PHLPP2 dephosphorylates AKT1, and both PHLPP1 and PHLPP2 dephosphorylate AKT3.

7.2.3 AKT Signaling and Metabolism

The activation of AKT in a cell greatly enhances the expression or translocation of GLUTs in the plasma membrane. This is followed by glucose transportation into the cell, the starting point of glucose metabolism. If AKT cannot be activated, then the cell suffers from energy depletion (due to the shortage of glucose to generate ATP) and activates its stress response pathway. If AKT becomes constitutively activated, then glucose uptake becomes uncontrolled, which constitutes a selection pressure for uncontrolled cell proliferation — the defining characteristic of cancer. The uncontrolled glucose flux would cause the cell to depend more and more on glycolysis to process glucose followed by lactic acid fermentation, even under the condition of abundant oxygen (aerobic glycolysis). Indeed, the citric acid cycle and oxidative phosphorylation are considerably slower than glycolysis in catabolizing glucose, which might make it more difficult to deal with the glucose influx. The PI3K-AKT-TOR pathway appears to be relevant to the Warburg effect (rapidly proliferating cells such as cancer cells tend to rely more on aerobic glycolysis for ATP generation).

In liver and muscle, the conversion of glucose to glycogen (glycogenesis) provides an energy store that, although not as efficient as triglycerides, can be rapidly mobilized (glycogenolysis) for a sudden surge in demand. AKT activation promotes glycogenesis by inhibiting GSK3β so that the enzyme glycogen synthase can be activated. The activation of AKT stimulates the phosphorylation of the Forkhead box O1 (FOXO1), a transcription factor that controls gluconeogenesis. The phosphorylation of FOXO1 reduces its entry into the nucleus, reduces the expression of key gluconeogenic genes, such as phosphoenolpyruvate carboxykinase (PEPCK), and reduces the net glucose output from the liver.

7.2.4 AKT Signaling and Whole-Body Glucose Homeostasis

Glucose homeostasis depends on tissue cells that remove glucose from the blood and thus contribute to the return of the plasma glucose concentration to the basal level. Muscle and fat cells are especially important, because they constitute most of the body mass and are thus the main source of glucose removal. The activation of the AKT in these cells induces a rapid relocation of GLUT4 from intracellular vesicles to the plasma membrane. When insulin becomes sufficiently high, the vesicles translocate to the plasma membrane and fuse with the plasma membrane, whereby GLUT4 is inserted onto the plasma membrane and becomes available for transporting glucose.

In brief, the majority of tissue cells rely on insulin-mediated AKT activation for glucose uptake.

In Chapter 3, we have identified bistable insulin response as the optimal strategy to satisfy both postabsorptive and fasting requirements. It is thus interesting to know whether or not AKT is activated in a bistable manner by insulin. This question will be answered in the next chapter in the context of analyzing the entire PI3K-AKT-TOR pathway. It turns out that the pathway works like a hysteretic, overflow valve to regulate plasma glucose concentration, and insulin indeed switches ON/OFF AKT activation in a bistable manner.

7.3 TARGET OF RAPAMYCIN (TOR)

The target of rapamycin (TOR) is a serine/threonine protein kinase that senses energy, nutrients, stress, and growth factors and integrates them to execute cell growth and division. Genetic and biochemical approaches in yeast and mammals led to the discovery of TOR as the target of the immunosuppressant rapamycin, a macrolide that is produced by a soil bacterium that is found on Easter Island. In the following, we use mTOR to represent TOR in mammalian organisms and leave TOR without a prefix for non-mammalian organisms. mTOR is the catalytic subunit of two distinct complexes called mTOR complex 1 (mTORC1) and mTOR complex 2 (mTORC2).

7.3.1 mTOR Complex 1

mTORC1 is mTOR outfits in RAPTOR (regulatory-associated protein of mTOR), which functions as a scaffold for assembling the complexes and for binding substrates and regulators. Other components of mTORC1 include a negative regulator PRAs40 (40 kDa Pro-rich AKT substrate; also known as AKT1s1), mLST8 (mammalian lethal with sEC13 protein 8; also known as GβL), and the recently identified DEPTOR (DEP domain-containing mTOR-interacting protein). Note that mLST8 and DEPTOR function as positive and negative regulators, respectively; the two are also present in mTORC2.

The mTORC1 substrates S6 kinase 1 (S6K1) and eIF4E binding protein 1 (4E-BP1) associate with mRNAs to regulate translation initiation and progression, whereby the rate of protein synthesis can be controlled. Upon phosphorylation by mTORC1, 4E-BP1 dissociates from eukaryotic translation initiation factor 4E (eIF4E), allowing eIF4E to recruit the translation initiation factor eIF4G to the 5′ end of most mRNAs; S6K1 promotes mRNA translation by phosphorylating or binding multiple proteins (e.g., eukaryotic elongation factor 2 kinase (eEF2K)), which collectively affect translation initiation and elongation.

mTORC1 positively regulates ribosome biogenesis (the synthesis of ribosomal RNAs (rRNAs) and ribosomal proteins), a highly energy-intensive process that is tightly coupled to the energetic status of the cell. mTORC1 upregulates the transcriptional activity of the rRNA polymerase RNA polymerase I (RNAPI) through S6K1 kinase. mTORC1 also controls the activity of several transcription factors that are implicated in lipid synthesis and mitochondrial metabolism.

mTORC1 actively suppresses autophagy, the controlled self-degradation of damaged, redundant, or even dangerous cellular components. There are two kinds of autophagy: microautophagy (of individual proteins) and macroautophagy (of entire organelles). Autophagy is a key process in providing substrates for energy production during periods of low extracellular nutrients.

7.3.2 mTORC1 Regulators

mTORC1 integrates four major regulatory signals: nutrients, growth factors, energy, and stress.

7.3.2.1 Nutrients

Amino acids are necessary for mTORC1 activation. In vertebrates, amino acids are largely sensed intracellularly rather than at the plasma membrane. The identity of the amino acid sensor is undetermined but there are several candidates, including the sterile 20 (STE20) family kinase mitogen-activated protein kinase kinase kinase kinase 3 (MAP4K3) and Class III PI3Ks (also known as Vps34).

The Rag family of small GTPases is arguably the strongest link between amino acids and mTORC1. Rag GTPases are heterodimers of either RAGA or RAGB with either RAGC or RAGD; the two members of the heterodimer have opposite nucleotide loading states. Rag GTPases are inactive when RAGA or RAGB is loaded with GDP and RAGC or RAGD is loaded with GTP. Rag GTPases become active when a switch occurs, in which RAGA or RAGB is loaded with GTP and RAGC or RAGD is loaded with GDP. Without (with) amino acids, the Rag GTPases are inactive (active). Rag GTPases reside in late endosomes and lysosomes. Therefore, the active Rag GTPase, by interacting with RAPTOR, causes mTORC1 to cluster onto the surface of late endosomes and lysosomes. This relocalization provides the opportunity for mTORC1 to interact with the small GTPase RHEB (Ras homologue enriched in the brain). In the absence of growth factors, RHEB is loaded with GDP and cannot activate mTORC1. With certain growth factors, RHEB is loaded with GTP and can thus activate mTORC1.

7.3.2.2 Growth Factors

For unicellular organisms, cell growth does not need much coordination, because each cell is an independent individual. In contrast, multicellular organisms rely on long-range communication to coordinate the distribution of nutrients and the growth of cell populations throughout the body. The communication is mediated by growth factors such as insulin and IGF1. And mTOR has to be wired into the PI3K-AKT pathway so that the growth signal, which carries the whole-body information, can be delicately relayed to mTOR. Binding of insulin or IGF1 to their corresponding receptors activates PI3K and then AKT. In turn, AKT phosphorylates a large protein TSC2 (also known as tuberin), that, together with TSC1 (also known as hamartin), forms the TSC1–TSC2 complex, which serves as a GTPase activating protein (GAP) for RHEB. In the absence of growth factors, TSC1–TSC2 is active and performs its GTPase function. Consequently, RHEB is loaded with GDP and cannot activate mTORC1. In the presence of growth factors, the activated AKT phosphorylates TSC2 and disables TSC2's GAP activity for RHEB, thus promoting mTORC1 activation. AKT also phosphorylates PRAS40, which causes it to bind to 14-3-3 proteins and prevents it from inhibiting mTORC1.

7.3.2.3 Energy

Because macromolecular synthesis (driven by mTORC1) is energy consuming, it is important that mTOR activity is under the control of energy. Otherwise, the starving cells would still synthesize proteins even when the energy is reduced below the basal level. Glycolysis and oxidative respiration convert nutrients into ATP. Upon nutrient deprivation, the intracellular ATP level drops quickly. mTORC1 actually senses the ratio AMP/ATP through AMP-activated protein kinase (AMPK). When the ratio of AMP to ATP increases, AMPK phosphorylates TSC2, possibly stimulating the GAP activity of TSC1–TSC2 toward RHEB to inhibit mTORC1 signaling. Moreover, AMPK phosphorylates RAPTOR, causing it to bind 14-3-3 proteins, which leads to the inhibition of mTORC1 through allosteric mechanisms.

7.3.2.4 Stress

Stress is closely related to energy. Many stressors affect ATP levels and thus may regulate mTOR through AMPK. During hypoxia, for example, the low oxygen concentration significantly impairs oxidative phosphorylation in the mitochondria, leading to low ATP levels and the activation of AMPK. Hypoxia also affects mTORC1 by inducing the expression of REDD1 (regulated in development and DNA damage response 1; also known as DDIT4), whose protein products suppress mTORC1 by promoting the assembly of TSC1–TSC2. There are other stressors that do not primarily impinge on the cellular energy signal through AMPK. For example, DNA damage leads to mTORC1 inhibition through the TP53-dependent upregulation of AMPK. TP53 targets sestrin 1 and sestrin 2, which are implicated in the DNA damage response. Sestrins potently activate AMPK and inhibit mTOR activity, which depends on TP53 but not energy.

7.3.3 mTOR Complex 2

mTORC2 is mTOR outfits in RICTOR (rapamycin-insensitive companion of mTOR). Other components include PROTOR1&2 (protein observed with RICTOR 1&2), which are likely to help complex assembly, mSIn1 (mammalian stress-activated map kinase-interacting protein 1, also known as mAPKAP1), which may target mTORC2 to membranes, mLST8 (which is the same molecule as in mTORC1), and DEPTOR (which is the same molecule as in mTORC1).

mTORC2 participates in key cellular processes such as cell cycle progression, anabolism, and cell survival. As already mentioned above, mTORC2 phosphorylates AKT at S473, which primes AKT for further phosphorylation at T308. Together, these two phosphorylation events cause the full activation of AKT. Besides AKT, mTORC2 phosphorylates many AGC kinase family members, such as serum- and glucocorticoid-regulated kinase (SGK) and protein kinase C (PKC), which regulate cell survival, cell cycle progression, and anabolism.

As for the upstream activators of mTORC2, growth factors likely play a role. Because AKT, SGK, and PKC respond to different growth factors, there may be multiple upstream regulators of mTORC2. To achieve signaling specificity, one possible solution may come from the existence of multiple isoforms of mSIN1. Three out of the five known splice variants of mSIN1 can be part of mTORC2, effectively defining three distinct complexes, two of which are regulated by insulin. Therefore, mSIN1 may function as an adaptor between mTORC2 and specific growth factor receptors.

7.3.4 Rapamycin and Its Effects on mTOR

In yeast and mammals, rapamycin inhibits the ability of TORC1, but not TORC2, to phosphorylate its substrates. Rapamycin binds the small protein FKBP12 (12-kDa FK506-binding protein; also known as PPIase FKBP1A) and, in turn, rapamycin–FKBP12 binds and inhibits RAPTOR-bound, but not RICTOR-bound, TOR. Rapamycin might inhibit TORC1 by dissociating RAPTOR from TOR, thus preventing the access of TOR to some substrates. Complicating this picture, prolonged treatment with rapamycin can inhibit mTORC2 in a subset of tissues and cell lines. This effect may involve a progressive sequestration of the cellular pool of mTOR in a complex with rapamycin–FKBP12, thus making it unavailable for assembly into mTORC2.

Rapamycin is an important cancer drug target, because it is naturally occurring. However, clinical trials thus far have generated only limited success. The results were unpredictable and the patients often suffered from rebound. The primary reason might be that rapamycin blocks the mTORC1 $\rightarrow$ S6K $\dashv$ IRS negative feedback loop. The inhibition of the negative feedback loop leads to a severe upregulation of PI3K signaling and provides important pro-survival and proliferative signals through AKT. Moreover, rapamycin does not fully inhibit all the functions of mTORC1: it only affects 4E-BP1 phosphorylation transiently and partially.

7.3.5 mTOR Signaling and Cell Growth

Cell growth is intimately linked to the availability of nutrients and energy. Under the conditions of nutrient starvation and low energy, cells rely on a complex set of programs to avoid energy imbalance and death: the suppression of macromolecular synthesis during fasting, the increase of the recycling of "aged" proteins and organelles to provide an internal source of metabolites, and the stoping or slowing down of proliferation.

The appearance of TOR (target of rapamycin) in unicellular organisms (e.g., *Saccharomyces cerevisiae*) enabled them to sense nutrient availability and to promote growth in favorable environmental conditions. The emergence of multicellularity necessitates a more complex network of molecular interactions (the PI3K-AKT-TOR pathway) to regulate growth and homeostasis, but TOR remains an important node.

Due to the intermittency of food intake, multicellular organisms need to store energy after feeding so that it can be used during periods of shortage. The requirement of keeping nutrient levels in the bloodstream within a narrow physiological range also necessitates this storage-mobilization mode. These mechanisms largely depend on mTOR regulation. During fasting, mTOR must be suppressed so that the limited cellular energy can be preserved for more essential functions. When nutrients and growth factors are available, mTOR is activated to drive anabolism (which consumes energy) and energy storage. However, chronic overfeeding can lead to an excess of mTOR activation and metabolic derangements.

7.3.5.1 Fasting and Starvation

During fasting and starvation, plasma glucose and amino acid levels decrease, causing a drop in circulating insulin. Within the cell, the AMP to ATP ratio increases due to the dwindling glucose availability. These factors place a brake on mTOR and inhibit energy-intensive macromolecular biosynthesis. Also, the cell begins to reutilize the constituents of cytoplasm and organelles (autophagy). For example, mitophagy (the autophagic degradation of mitochondria) provides an immediate source of energy at the expense of nutrient-intensive, long-term ATP production. The effect of autophagy can be so dramatic that the murine liver shrinks to around one-third of its original size after a 24-h fasting, due to the recycling of mitochondria, cytoplasmic proteins, and stored glycogen. Furthermore, white adipose tissue and hepatocytes mobilize lipid stores, converting them into FFAs that are utilized by the liver and muscle through β oxidation. Autophagosomes can help to sequester lipid droplets and break them down into FFAs. Autophagy also mediates massive protein breakdown in the muscle, releasing amino acids into the bloodstream to be converted to glucose in the liver. The released nutrients (amino acids, glucose, and other metabolites) feed back onto mTORC1, causing its partial reactivation.

Mitophagy is only one aspect of mitochondrial changes induced by mTOR inhibition. We know that mTORC1 promotes biosynthesis through many actions, including the inhibition of the protein 4E-BP1. Therefore, starvation-mediated mTORC1 inhibition would activate 4E-BP1, which in turn inhibits biosynthesis. Despite the global inhibition of translation, the translation of mRNAs encoding the mitochondrial respiratory chain is enhanced. Therefore, mTOR inhibition during starvation acts on mitochondria at three different levels. First, a subset of the existing mitochondria is eliminated (mitophagy). Second, the synthesis of new mitochondria is reduced. Finally, the efficiency of existing mitochondria is enhanced through the 4E-BP1 translational program.

7.3.5.2 Overfeeding and Insulin Resistance

In the evolutionary time scale, nutrient abundance is only recent, and nutrient scarcity was prevalent during mammalian evolution. Therefore, natural selection might favor the storage-mobilization mode of energy metabolism, i.e., the storage of energy in anticipation of periods of shortage. In this event, the constant availability of food in large quantity, characteristic of many current populations, may lead to chronic mTORC1 activation, which contributes to obesity by mediating excess fat deposition in WAT. In addition, chronic mTORC1 activation contributes to ectopic fat deposition in liver and muscle, which has a role in the insurgence of insulin resistance. Moreover, mTORC1 hyperactivation triggers an S6K1-dependent negative feedback loop: activated S6K1 dampens the function of IRS, which in turn weakens the level of AKT activation and thus increases insulin resistance. This negative feedback is paradoxical: how could mTORC1 be constitutively active in an insulin-resistance state, given the importance of insulin in mTOR activation? This paradox can be explained if excess nutrients can compensate for the effect of insulin resistance in mTOR activation. Indeed, it is known that chronically high blood levels of amino acids, as seen in obesity, will keep mTORC1 at work. The role of negative feedback can be better understood in the context of the entire PI3K-AKT-mTOR pathway, with mathematical modeling as a tool of analysis (see Chapter 9).

mTORC1 coordinates food intake (systemic control) with energy storage and expenditure (in peripheral tissues). The stimulation of mTORC1 (by, e.g., leucine) in the hypothalamus induces satiety. Conversely, the inhibition of mTORC1 (by, e.g., rapamycin injection) in the hypothalamus increases food intake. This multilevel regulation explains the profound consequences that deregulated mTOR signaling exerts on human metabolism.

7.3.5.3 Adipogenesis and Lipogenesis

mTOR signaling plays a fundamental role in the maintenance of fat cells and the formation of adipose tissues, the most important energy storage site in mammals. Many downstream effectors of mTORC1 are involved in the control of adipogenesis. S6K1 regulates the commitment of embryonic stem cells to adipogenic progenitors by regulating the expression of early adipogenic transcription factors; 4E-BPs control the terminal differentiation of adipocytes through the translational control of the master regulator of adipogenesis, PPARγ. In addition, mTORC1 may affect PPARγ activity by activating the transcription factor sterol regulatory element-binding protein 1c (sREBP1c) that, in turn, mediates the production of a PPARγ-activating ligand. Besides adipogenesis, mTORC1 also mediates lipogenesis, which involves the synthesis of triglycerides stored in white adipose tissue.

mTORC2 is also important in lipid biogenesis, because it contributes the full activation of the master regulator AKT through the phosphorylation at S473. AKT activation leads to the induction of PPARγ as a result of its ability to activate mTORC1 and to directly inhibit FOXO1, which usually antagonizes PPARγ activity.

mTORC1 is highly active in the tissues of obese rodents, a disease state characterized by the expansion of adipose tissue. Elevated circulating insulin, proinflammatory signals, and nutrients (branch-chain amino acids and glucose) are driving forces that promote mTORC1 activity in the obese state. mTORC1 also promotes insulin resistance in adipose tissue through the S6K1-mediated negative feedback inhibition of IRS signaling. The reduction in the action of insulin in adipose tissue probably exacerbates systemic insulin resistance by promoting FFA release by adipocytes, ectopic fat deposition, and lipotoxicity. The high rate of protein synthesis associated with mTORC1 activation may also exacerbate insulin resistance by promoting endoplasmic reticulum stress and the unfolded protein response. Endoplasmic reticulum stress is a condition that prevails in enlarged adipocytes where it impairs insulin signaling through the destabilization of IRS by c-Jun N-terminal kinase.

7.3.6 MTOR Signaling and Cancer

The mutations of negative mTORC1 regulators (e.g., TSC1–TSC2 and LKB1) can drive tumorigenesis, as demonstrated by familial cancer syndromes arising from these mutations. For the downstream target of mTORC1, 4E-BP1 represents a prominent oncoprotein. The inactivation of 4E-BP1 may directly contribute to the growth of sporadic cancers. This is because eIF4E, which is inhibited by 4E-BP1, preferentially drives the translation of mRNAs for pro-tumorigenic genes, including cell cycle regulators. Therefore, the inhibition of 4E-BPs by mTORC1 drives cell cycle progression but not cell growth. In addition, eIF4E promotes cell survival in in vivo mouse models of lymphoma by upregulating the translation of the anti-apoptotic protein myeloid leukemia cell differentiation 1 (MCL1). In addition, mTORC1 indirectly upregulates the enzyme fatty acid synthase (FAS), a transcriptional target of SREBP1 involved in lipid biogenesis that favors the rapid proliferation of cancer cells. Constitutive mTORC1 activation may favor carcinogenesis by suppressing autophagy, a biological process that helps to suppress tumors. Constitutive mTORC1 activation enhances the expression of hypoxia inducible factor 1 (HIF1) and promotes angiogenesis, which favors tumor growth by supplying them with nutrients and oxygen.

mTORC2 also favors tumorigenesis due to its role in AKT activation, because AKT promotes proliferation, survival, and nutrient uptake in cancer cells and their metastasis. In fact, RICTOR is required for the growth of tumor cell lines and prostate tumors in PTEN-deficient mice.

7.4 PATHWAY DESCRIPTION

In the above sections, the three key nodes of the pathway, PI3K, AKT, and TOR, were described in detail, as well as their local connections. In this section, the components of the pathway are assembled to form a (simplified) network of interactions (Figure 7.1). Although many growth factors can activate the pathway, in the following only insulin is used as an example to elucidate the functioning of the PI3K-AKT-TOR pathway.

Insulin binds to the insulin receptor and leads to the phosphorylation of IRS, which contains a conserved pleckstrin homology domain located at its N terminus that anchors it to membrane phosphoinositides in close proximity to the insulin receptor. The pleckstrin homology domain is flanked by a phosphotyrosine-binding domain, which enables IRS to bind with the insulin receptor. The C-terminal region of IRS contains multiple tyrosine phosphorylation motifs that serve as docking sites for the p85α regulatory subunit of PI3K, whereby PI3K is activated.

PI3K targets many substrates, including PIP2, which is a minor phospholipid component of the plasma membrane of the cell and is in equilibrium with PIP2. Activation of PI3K shifts the PIP2:PIP3 equilibrium toward PIP3, which binds both AKT and PDK1 and recruits them to the plasma membrane. This provides the opportunity for PDK1 to phosphorylate AKT on T308. On the other hand, TORC2 phosphorylates AKT on S473, contributing to the full activation of AKT. Because AKT is a master regulator, its activation enables many biological functions, such as energy metabolism, cell growth, and proliferation. For example, the activated AKT promotes the translocation of GLUT4 from intracellular vesicles to the plasma membrane and thus enables massive glucose uptake.

The activated AKT phosphorylates the tuberous sclerosis complex 2 (TSC2) protein tuberin, and therefore inhibits the GTPase-activating protein (GAP) activity of the TSC1–TSC2 complex toward RHEB (small G protein Ras homologue enriched in the brain). This allows GTP-bound RHEB to accumulate and to activate the TORC1

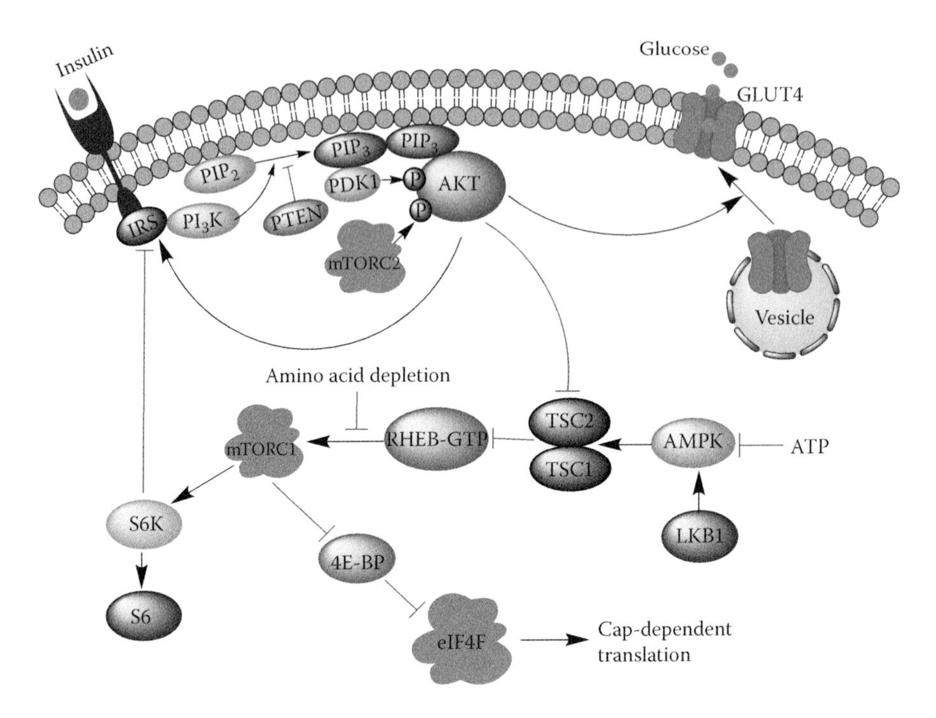

FIGURE 7.1 The structure of the PI3K-AKT-TOR pathway. It is a simplified network, with some intermediate molecules omitted.

complex under the condition of sufficient amino acid concentration. The activated TORC1 in turn mediates phosphorylation of 4E-BP1 and S6K, ultimately leading to increased protein synthesis.

Of note, the activated AKT regulates IRS both positively and negatively. On one hand, AKT directly interacts with IRS and enhances its activity (through phosphorylation of serine residues), which generates a positive feedback loop. On the other hand, AKT reduces IRS activity indirectly (through the route AKT $\rightarrow$ TORC1 $\rightarrow$ S6K $\dashv$ IRS), which forms a negative feedback loop. S6K phosphorylates Ser residues at the C terminus of IRS. This phosphorylation prevents insulin-stimulated Tyr phosphorylation of IRS-1 at this region and thus blocks its ability to bind PI3K. The phosphorylation also has other negative effects on IRS activity. It induces the dissociation of IRS proteins from the insulin receptor and promotes the release of the IRS proteins from intracellular complexes that maintain them in close proximity to the receptor. It induces degradation of IRS proteins and may even turn IRS proteins into inhibitors of the insulin receptor kinase. The negative feedback has many important biological functions. TORC1 carries information about intracellular nutrients, because it is a sensor of energy status and amino acid concentration. The negative feedback can thus inform the cell about how much it has taken. If the intracellular nutrient level has already been high (low), then the negative feedback tells the cell to absorb less (more) nutrients. Overactivity of the negative feedback, however, may lead to insulin resistance and other pathological consequences, due to the enhanced downregulation of insulin signaling.

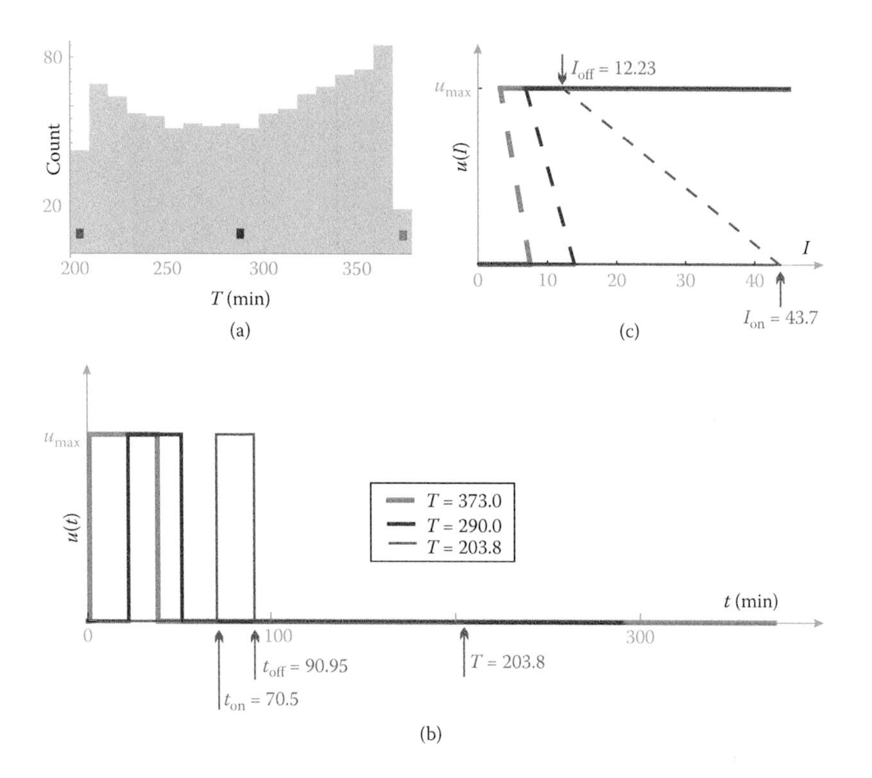

FIGURE 3.6 The set of locally optimal controls whose utilization rates are within the range $\eta \pm \Delta\eta$, where $\eta = 0.1$ and $\Delta\eta = 5 \times 10^{-5}$. (a) The distribution of control counts over the T values. (b) Three $u(t)$ are illustrated: the worst control (green, $T = 373$), an average control (blue, $T = 290$), and the globally optimal control (red, $T = 203.8$). (c) The three controls in the form of $u(I)$. They are obtained by following the conversion procedure in Fig. 3.2.

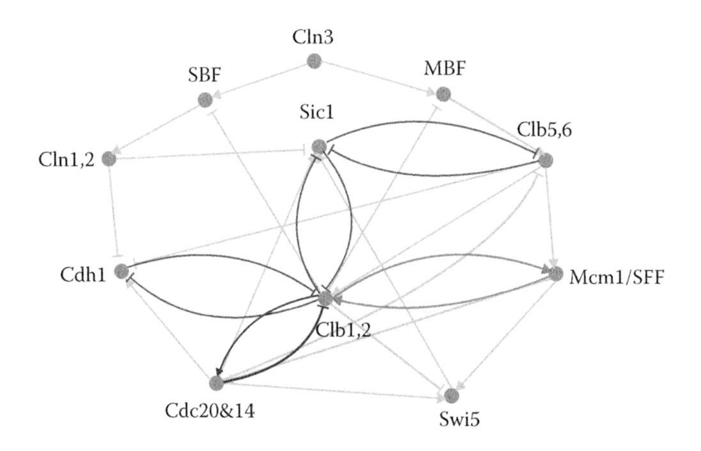

FIGURE 6.3 Feedback loops in the budding yeast cell cycle network. The negative feedback, positive feedback, and double negative feedback loops are colored in blue, green, and red, respectively.

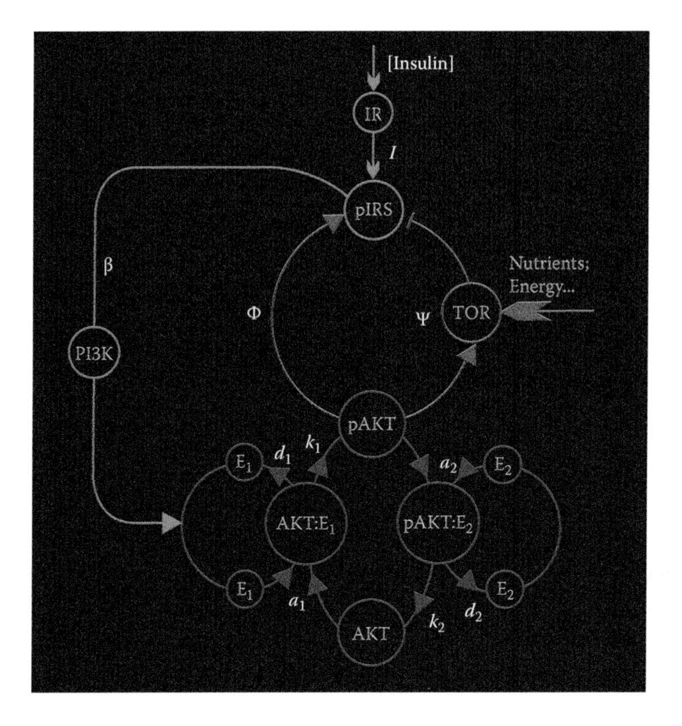

FIGURE 9.1 The backbone structure of the PI3K-AKT-TOR pathway. The pathway can be divided into three components: the input (colored in green), the output (colored in blue), and the feedback (colored in red).

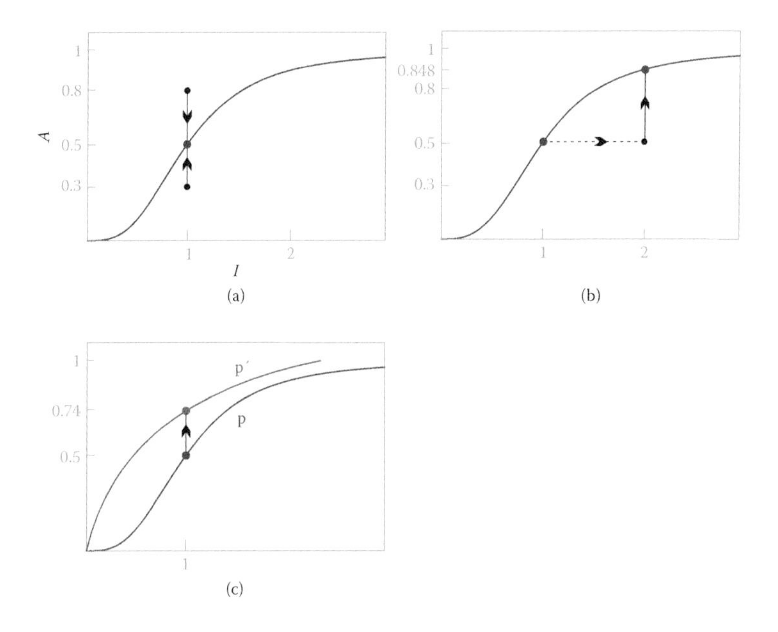

FIGURE 9.2 The response curve and the network state perturbation. (a) Random perturbations of the A value while keeping I fixed. (b) The perturbation of I to 2 causes the network state to change to a new steady state. (c) The change of the parameters makes the entire response curve change.

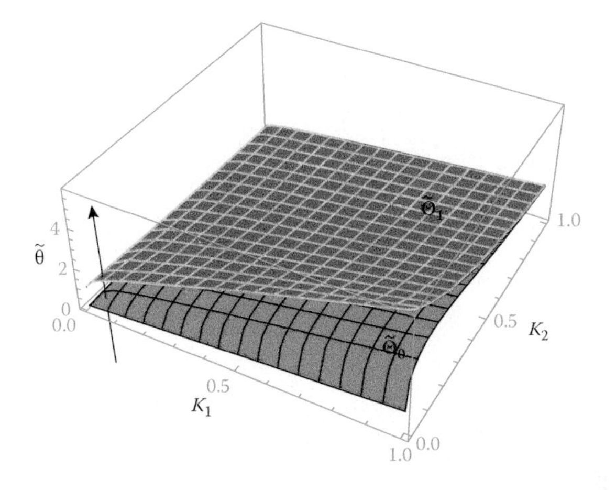

FIGURE 9.3 The space of $\tilde{\mathbf{p}} = \left(\tilde{\theta}, K_1, K_2\right)$. The surface $\tilde{\Theta}_1(K_1, K_2)$ (red) and the surface $\tilde{\Theta}_0(K_1, K_2)$ (green) divide the space into three regions, which from the top down are the irreversible switch, the toggle switch, and the monotone type.

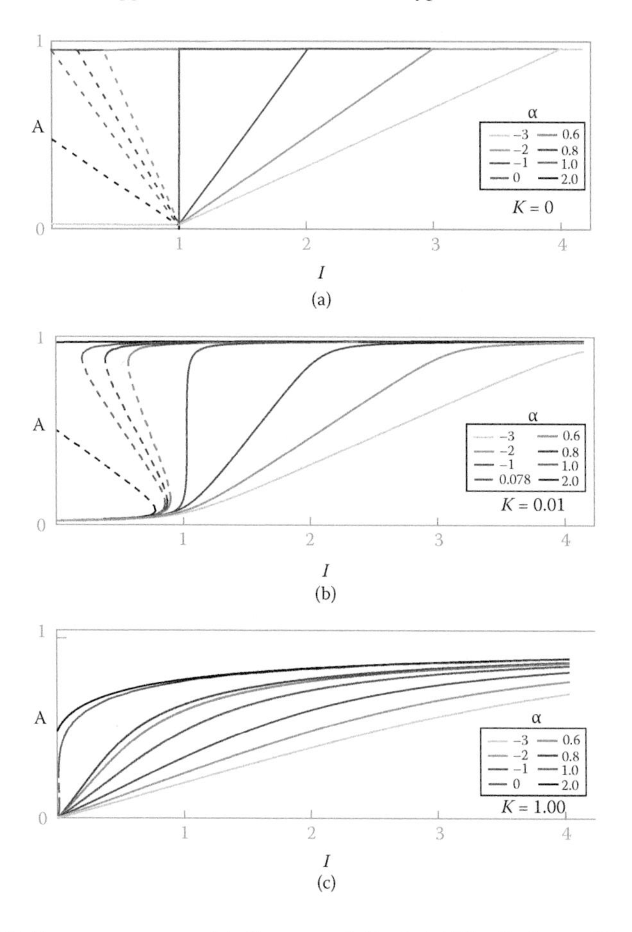

FIGURE 10.2 Response curves with $\beta = \gamma = 1$ fixed. (a) The limit condition $K = 0$. (b) The condition $K = 0.01$. (c) The condition $K = 1.00$.

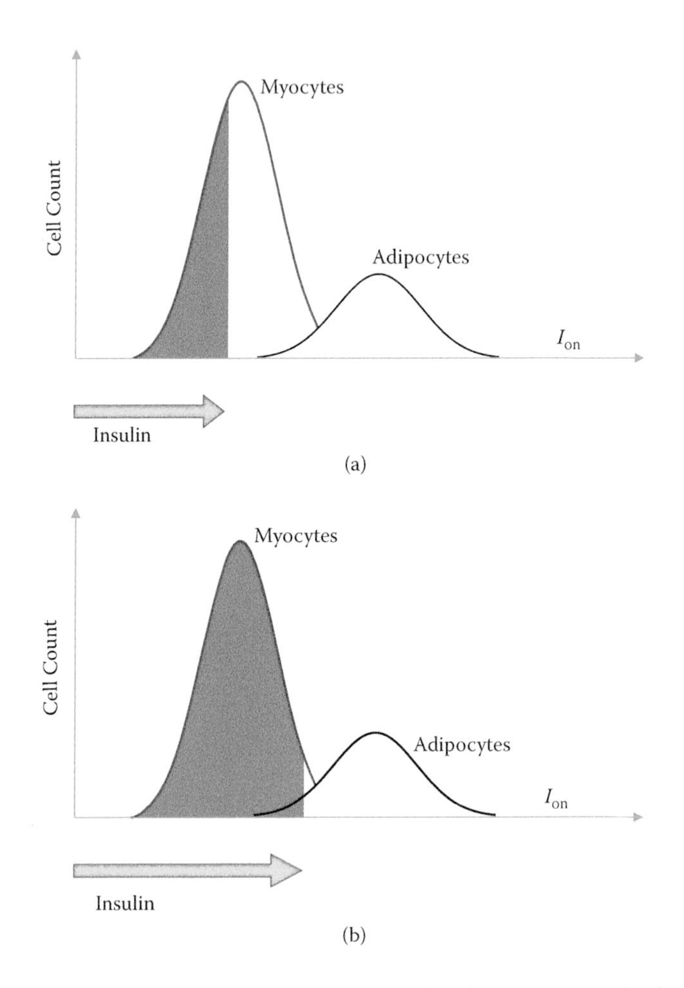

FIGURE 10.4 Myocytes are heterogeneous in terms of their I_{on} values; their distribution is represented by the red curve. Adipocytes are also heterogeneous in terms of their I_{on} values; their distribution is represented by the black curve. The yellow arrow represents the peak insulin level. (a) The peak insulin level is smaller than all the I_{on} values of adipocytes; thus no adipocytes are activated. The portion of activated cells is illustrated in green. (b) The peak insulin level is larger than the I_{on} values of a small portion of adipocytes. These adipocytes are thus activated (indicated by the small region colored in green underneath the black curve).

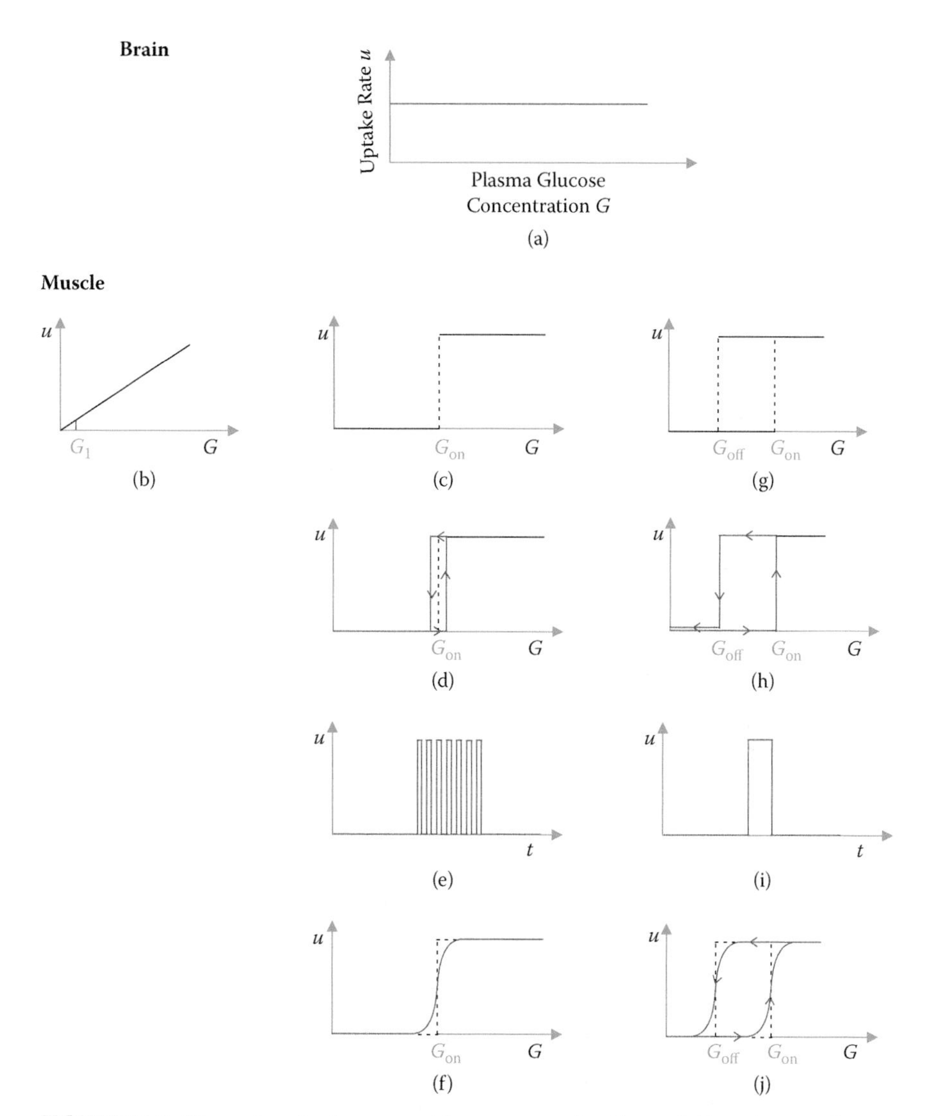

FIGURE 11.1 The rate of glucose uptake u as a function of the plasma glucose concentration G. (a) The brain has a relatively constant rate of glucose uptake. (b) Putative pattern of muscular glucose uptake: a gradual increase as G increases. (c) Putative pattern of muscular glucose uptake: a step function that switches at G_{on}. (d) A reproduction of (c), with the clattering phenomenon highlighted in red. (e) Putative muscular glucose uptake as a function of time, according to (c). (f) A modification of (c), with the effect of cell heterogeneity considered. (g) Putative pattern of muscular glucose uptake: a toggle switch that turns on at G_{on} and turns off at G_{off}. (h) A reproduction of (g), which highlights in red the sensitive and robust response. (i) Putative muscular glucose uptake as a function of time, according to (g). (j) A modification of (g), with the effect of cell heterogeneity considered.

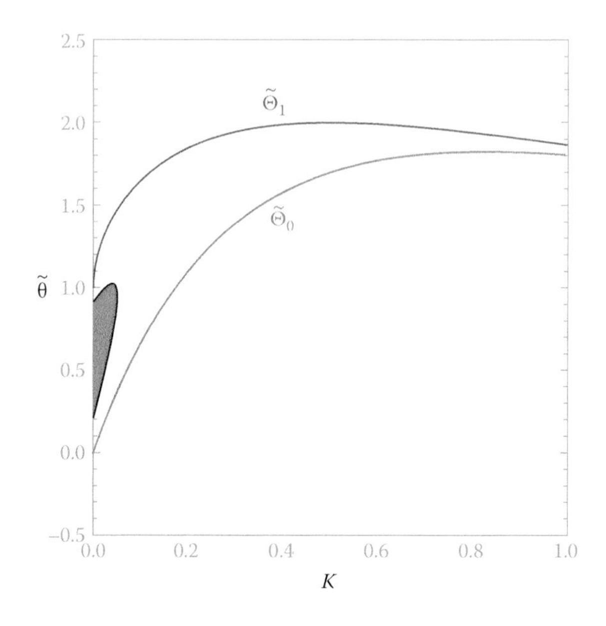

FIGURE 11.2 The phase diagram K versus $\tilde{\theta}$, where the rough region of normality is colored in green.

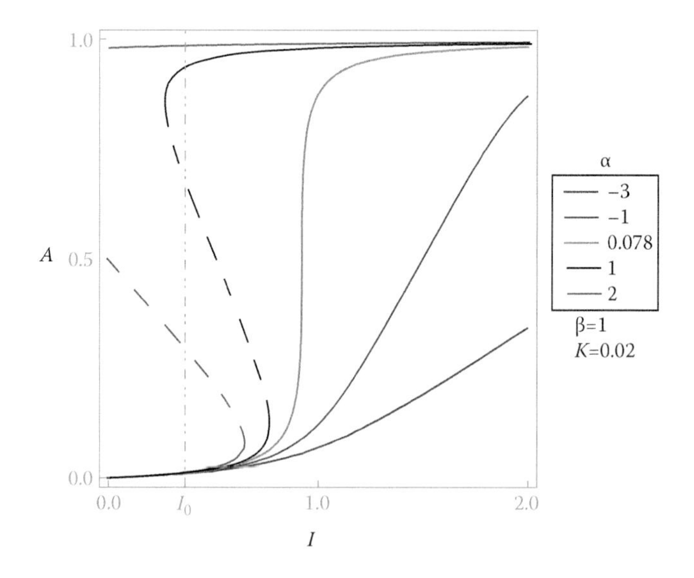

FIGURE 12.1 Response curves $A(I)$ that may correspond to physiologic/pathologic phenotypes. These curves are parameterized by the same values of β and K, but different α values. I_0, indicated by the dash-dot-dot line, represents the basal level of the growth factor I.

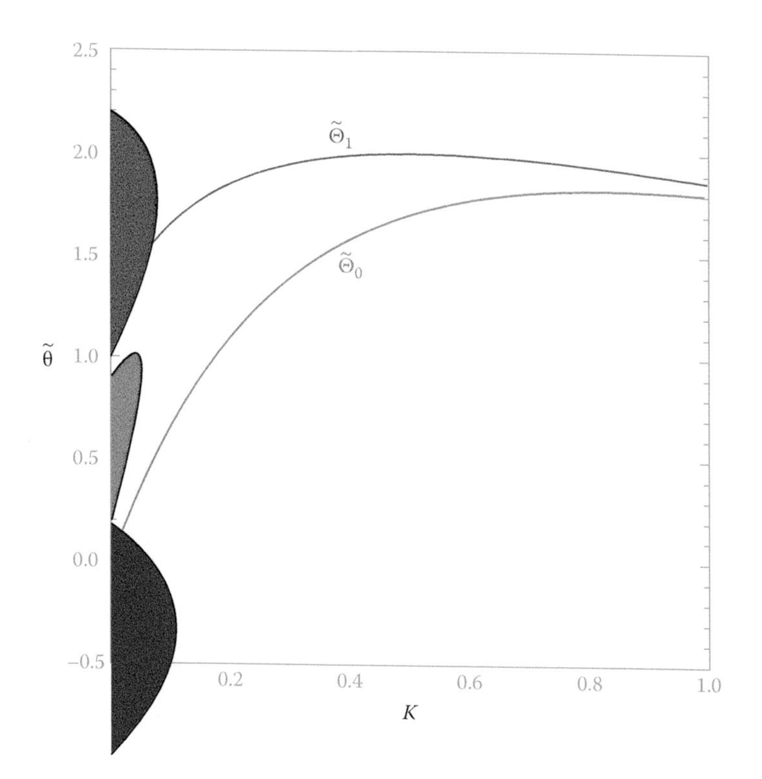

FIGURE 12.2 The phase diagram K versus $\tilde{\theta}$, where the rough regions of normality, cancer, and diabetes are colored in green, red, and blue, respectively.

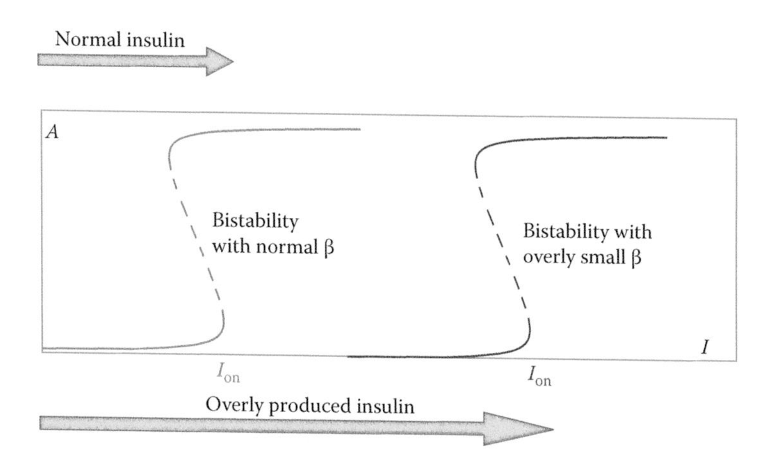

FIGURE 12.3 Insulin resistance caused by the decrease of β. The green curve corresponds to normal β, which allows for a normal amount of insulin for AKT activation. The blue curve corresponds to overly small β, which requires overproduction of insulin to activate AKT.

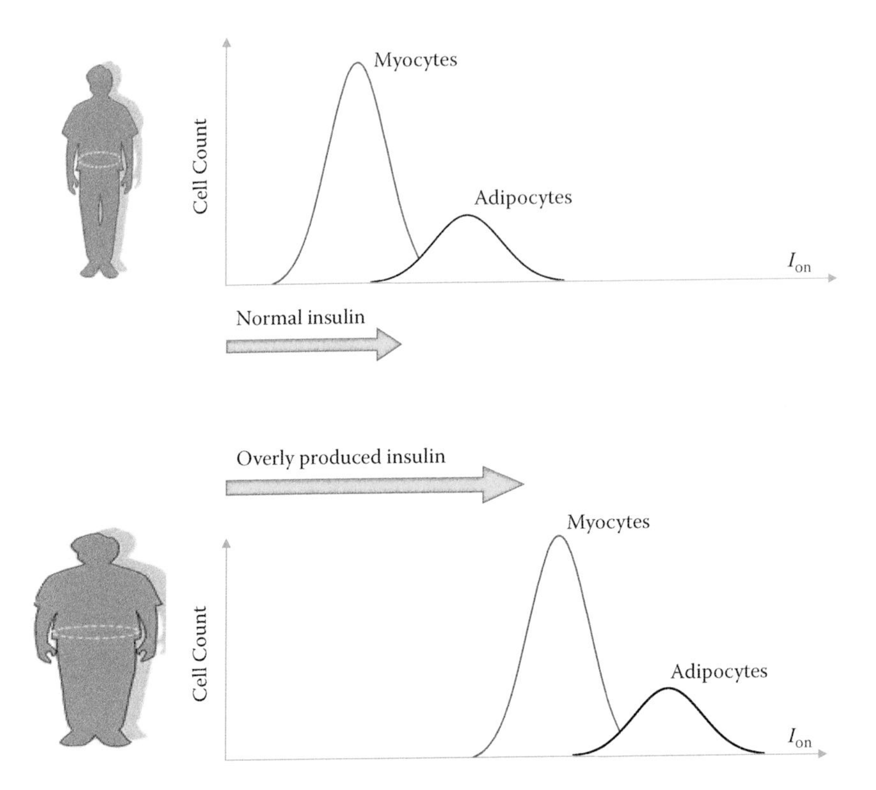

FIGURE 12.4 The distribution of myocytes and adipocytes over their I_{on} values, for a person in both the lean (upper panel) and obese (lower panel) states. As the person becomes obese, the I_{on} values shift to the right, for both myocytes and adipocytes. The pancreas has to secrete more insulin to trigger glucose uptake.

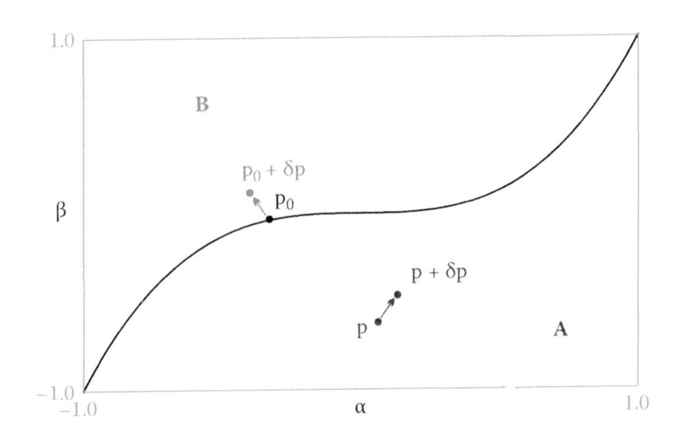

FIGURE C.1 The two-dimensional space $\mathbf{p} = [\alpha, \beta]$ that is divided into two regions A and B.

Part III

Mathematical Analysis of Complex Diseases

Our lives are threatened by complex diseases, many of which are untreatable. What makes matters even worse are the three epidemics (cancer, obesity, and diabetes), the incidence of which has escalated in recent years. In the United States, about two-thirds of the population is overweight; about one-third of children born after the year 2000 will develop diabetes; about one-third of women and half of men will develop cancer at some point during their lifetimes. In the last few decades, the number of people with diabetes has more than doubled globally, making the combination of type 2 diabetes and obesity the largest epidemic the world has yet faced. These threats urge the development of novel concepts and theories to explore the nature of these diseases. In this book, a holistic perspective (multiple diseases are considered as qualitatively different but interconnected entities that bifurcate from normal) is proposed, from which a deep understanding of diseases can be developed by applying mathematical techniques such as singularity analysis.

Let's begin with a new theory of diseases called *normality deviation*. In the theory, a complex disease arises from an infinitesimal deviation from normality, namely, a small aberration that preludes the progression of a specific disease. Because the deviation can occur in different ways, the target of study is actually not a disease in isolation, but a collection of distinct but related diseases. The crucial idea here is infinitesimal, which allows one to use powers from mathematics (calculus, singularity analysis, topological equivalence, etc.) to unravel the structure around normality, which in turn leads to the classification of diseases. This infinitesimal approach only applies to the early, asymptomatic phase of diseases. Nevertheless, the asymptomatic phase is the most important, during which the essential aberration has already occurred and will dictate the entire disease progression. The later phase is usually dominated by nonessential body reactions that manifest themselves as common symptoms. Putting these ideas together, a global picture emerges in which diseases deviate from normality over time, in ways that are small initially but eventually become dramatic.

Then, what causes the initial aberration? It is true that a gene mutation can kick off disease progression. However, genes and their products are not isolated, but are constituents of a network of interactions. It is the system-level response of the biomolecular network, rather than any individual molecules, that determines what is normal and what are aberrations. To reveal system-level control mechanisms, the biomolecular network needs to be modeled and analyzed mathematically, with the response curves of the system characterized. The response curves can be classified into different groups. Each group includes response curves of the same kind that are qualitatively different from other groups. The boundaries between two different groups (which is a set of curves that are close to both groups but belong to neither group) are what mathematicians call singularity. Being special, singularities are much easier to determine (by using analytical techniques so that numerical computations can be simplified). Once the singularities are determined, the landscape has already been illuminated. One may use an analogy with drawing an object. By tracing out the boundary between the object and air, one needs not depict every point of the object. This is why considering the global picture of related diseases can actually facilitate analysis instead of perplexing the situation.

Of course, it is impossible to include all the diseases in the same framework. It is also not necessary to do so. In this book, we focus on diseases related to energy metabolism and cell growth, which include cancer, diabetes, obesity, the biggest epidemics in human history. Because the PI3K-AKT-TOR pathway is the molecular circuit controlling metabolism and growth and because the pathway is well conserved during evolution, a mathematical model will be developed for the pathway so that the theory of normality deviation will have a molecular basis. Analysis of the mathematical model will help to unravel the control mechanisms of the PI3K-AKT-TOR pathway, including different types of response that underlie different disease progressions.

A complex disease is almost always multifactorial, involving many risk factors operating at different levels. This book does not aim for a comprehensive discussion of all the aspects of the selected diseases. Instead, it centers on metabolism and growth, which are important aspects of cancer, diabetes, and obesity. In fact, uncontrolled cell growth is the defining characteristic of cancer. Cancer metabolism (including the Warburg effect) was once the hottest field in cancer biology and is now undergoing a revival of research interest. Obesity and diabetes are metabolic diseases in the first place.

This book only considers the latency phase of the diseases, because the underlying theory depends on the concept of "infinitesimal deviation from normality." Nevertheless, aberrations that occur at the latency phase are essential and will persist throughout the disease progression. Understanding of these mechanisms provides valuable insights into later complications of these diseases.

Although only diseases related to growth and metabolism are considered in this book, the paradigm used in the book (the normality deviation theory and the singularity analysis approach) can be useful for studying other categories of diseases, i.e., diseases related in some other way.

8 Diseases Related to Metabolism

8.1 CANCER

Cancer is actually a broad group of various diseases, known as malignant neoplasm medically. Cancer cells grow and divide with little or no control, forming malignant tumors, and invade nearby parts of the body. The cancer may also spread to more distant parts of the body through the lymphatic system or bloodstream (metastasis). There are over 200 different known cancers that afflict humans.

Douglas Hanahan and Robert A. Weinberg proposed six hallmarks of cancer in 2000. Recently, two more hallmarks were added to the list [15]. These hallmarks provide a logical framework for understanding the remarkable diversity of neoplastic diseases. During the transformation from normal to neoplastic, the cells acquire a succession of these hallmark capabilities, which are distinctive and complementary traits the cells have to acquire in order to become tumorigenic and ultimately undergo metastatic dissemination. These hallmarks are described as follows.

8.1.1 Sustaining Proliferative Signaling

The growth and division of normal tissues are under both cell autonomous control and systemic control, thereby ensuring a homeostasis of cell number and thus maintenance of normal tissue architecture and function. The systemic control is usually mediated by growth factors that bind cell-surface receptors, typically containing intracellular tyrosine kinase domains. The latter proceed to emit signals via branched intracellular signaling pathways that regulate progression through cell growth (i.e., increases in cell size) as well as cell proliferation (i.e., increases in number). The growth factors controlling cell number and position within tissues appear to be transmitted in a temporally and spatially regulated fashion from one cell to its neighbors, but such paracrine signaling is difficult to access experimentally.

Cancer cells sustain proliferative signaling in many different ways. Cancer cells can either produce growth factors themselves (autocrine) or stimulate normal stromal cells to produce growth factors. Cancer cells may also upregulate the levels of receptor proteins so that they can be hyperresponsive to growth signals. Mathematical analysis in the next chapter will reveal another mechanism of sustained signaling — the constitutive activation of an oncoprotein upon transient growth factor stimulation. Although the mechanism may be a universal one and applicable to many signaling pathways, the following discussions are primarily based on the PI3K-AKT-TOR pathway. The production of growth factors (insulin, IGF, EGF, etc.) in the body is usually in a pulsatile, transient manner. Nevertheless, the transient production of a growth factor can generate a long-lasting effect — the constitutive activation of the PI3K-AKT-TOR pathway even while the growth factor itself has been withdrawn completely. Because activation of the pathway boosts cell growth in every aspect (from energy acquisition to macromolecular synthesis), the constitutive activation of the pathway makes uncontrolled cell growth and proliferation possible.

In the next chapter, the question will be answered as to why the PI3K-AKT-TOR pathway may be induced by growth factors to run out of control.

8.1.2 EVADING GROWTH SUPPRESSORS

Cancer cells must also circumvent powerful programs that negatively regulate cell proliferation. Some programs depend on the actions of tumor suppressor genes, including the two prototypical tumor suppressors that encode the RB (retinoblastoma-associated) and TP53 proteins. The two proteins integrate signals from diverse extracellular and intracellular sources to decide whether or not a cell should proceed through its growth-and-division cycle. While the two proteins' functions largely overlap, RB transduces growth-inhibitory signals that originate largely outside of the cell and TP53 receives inputs from stress and abnormality sensors that function within the cell's intracellular operating systems.

Contact inhibition is another mechanism used by normal cells to suppress uncontrolled growth. Cell-to-cell contacts formed by dense populations of normal cells can conduct signals to suppress further proliferation of the cells. Unfortunately, contact inhibition is abolished in various types of cancer in cell culture experiments, which may also occur during tumor growth in vivo. The genetic basis of contact inhibition is beginning to emerge. Merlin, the cytoplasmic product of the gene *NF2*, orchestrates contact inhibition via coupling cell-surface adhesion molecules (e.g., E-cadherin) to transmembrane receptor tyrosine kinases (e.g., the EGF receptor), whereby the adhesivity of cadherin-mediated cell-to-cell attachments can be strengthened. This action also sequesters growth factor receptors, thus limiting the efficient transmission of mitogenic signals. Another example is the LKB1 protein, which organizes epithelial structure and helps to maintain tissue integrity. LKB1 can overrule the mitogenic effects of the powerful *myc* oncogene when the latter is upregulated in organized, quiescent epithelial structures. When the gene *LKB1* is deleted or its expression is suppressed, epithelial cells become susceptible to *myc*-induced transformation. Note that LKB1 has another important function of preventing TORC1 hyperactivity, as already mentioned.

TGF-β is best known for its antiproliferative effects, and the corruption of the TGF-β pathway is now understood to cause uncontrolled cell proliferation. In addition to suppressing cell proliferation, the TGF-β pathway was found, in many late-stage tumors, to activate a cellular program, termed the epithelial-to-mesenchymal transition (EMT), that confers on cancer cells traits associated with high-grade malignancy.

8.1.3 RESISTING CELL DEATH

Programmed cell death (apoptosis) is a natural barrier to cancer development. Although many risk factors can kick off tumorigenesis, progression to malignancy is often inhibited because apoptosis is triggered in response to various physiologic stresses caused by deregulation and rapid growth: signaling imbalances resulting from elevated levels of oncogene signaling, DNA damage associated with hyperproliferation, and so on. The apoptotic machinery is composed of both an upstream regulator component and a downstream effector component. The regulator component can be divided into one circuit that receives and processes extracellular death-inducing signals (e.g., the Fas ligand/Fas receptor) and another circuit that senses and integrates a variety of intracellular signals. The effector component consists of a cascade of proteolysis involving effector caspases responsible for the execution phase of apoptosis, in which the cell is progressively disassembled and then consumed, both by its neighbors and by professional phagocytic cells. The apoptotic trigger that conveys signals between regulators and effectors is controlled by balancing proapoptotic proteins (Bax, Bak, cytochrome c) and antiapoptotic proteins (Bcl-2, Bcl-xL, Bcl-w, Mcl-1, A1). Bax and Bak are embedded in the mitochondrial outer membrane. When relieved of inhibition by their antiapoptotic relatives, Bax and Bak disrupt the integrity of the outer mitochondrial membrane, causing the release of cytochrome c and/or other proapoptotic signaling proteins. Cytochrome c then activates a cascade of caspases that act via their proteolytic activities to induce the multiple cellular changes associated with the apoptotic program.

Apoptosis is attenuated in those tumors that succeed in progressing to states of high-grade malignancy and resistance to therapy. Tumor cells evolve a variety of strategies to limit or circumvent apoptosis, most notable of which is the loss of TP53 tumor suppressor function, which eliminates a critical DNA damage sensor from the apoptosis-inducing circuitry. Alternatively, tumors may increase the expression of antiapoptotic regulators (Bcl-2, Bcl-xL) or of survival signals (Igf1/2) by downregulating proapoptotic factors (Bax, Bim, Puma) or by short-circuiting the extrinsic ligand-induced death pathway. The multiplicity of apoptosis-avoiding mechanisms presumably reflects the diversity of apoptosis-inducing signals that cancer cell populations encounter during their evolution to the malignant state.

8.1.4 Enabling Replicative Immortality

The emergence of a macroscopic tumor involves unlimited growth and replications. In contrast, normal cells only pass through a limited number of successive cell growth and division cycles, due to two distinct barriers to proliferation. The first is senescence, which is an irreversible entrance into a nonproliferative but viable state. The second is crisis, which involves cell death. When cells are proliferating, repeated cycles of division lead first to induction of senescence and then, for those cells that succeed in circumventing this barrier, to a crisis phase, in which most of the cells die. On rare occasions, cells emerge from a population in crisis and exhibit unlimited replicative potential. This transition has been termed immortalization, a trait that most established cell lines possess by virtue of their ability to proliferate without evidence of senescence and crisis.

Unlimited proliferation is an extraordinary capability that depends on telomeres, which are multiple tandem hexanucleotide repeats protecting the ends of chromosomes. Telomeres shorten progressively in normal cells and eventually lose the ability to protect the ends of chromosomal DNAs from end-to-end fusions; such fusions generate unstable dicentric chromosomes that threaten cell viability. Therefore, the length of telomeric DNA dictates how many successive cell generations its progeny can pass through before the erosion of telomeres and the entrance into crisis. Telomerase, the specialized DNA polymerase that adds telomere repeat segments to the ends of telomeric DNA, is almost absent in nonimmortalized cells but expressed at functionally significant levels in the vast majority (90%) of spontaneously immortalized cells, including human cancer cells. By extending telomeric DNA, telomerase is able to counter the progressive telomere erosion that would otherwise occur in its absence. Due to the functioning of telomerase, cancer cells can maintain telomeric DNA at lengths sufficient to avoid triggering senescence or apoptosis, achieved most commonly by upregulating the expression of telomerase or, less frequently, via an alternative recombination-based telomere maintenance mechanism. Hence, telomere shortening is thought to be a clocking device that determines the limited replicative potential of normal cells, which must be overcome by cancer cells.

8.1.5 Inducing Angiogenesis

Tumor cells need a large quantity of nutrients and oxygen to sustain their rapid divisions. By the process of angiogenesis, tumors develop neovasculature so that their nutrient and oxygen demands can be met, as well as the requirements of evacuating metabolic wastes and carbon dioxide. During tumorigenesis, an "angiogenic switch" is almost always activated and remains on, causing new vessels to be sprouted continually, which helps sustain expanding neoplastic growths. In contrast, the vasculature of normal cells becomes largely quiescent after their development during embryogenesis.

The angiogenic switch is governed by angiogenesis inducers (e.g., vascular endothelial growth factor-A (VEGF-A), fibroblast growth factor (FGF)), and inhibitors (e.g., thrombospondin-1 (TSP-1)). VEGF gene expression can by upregulated both by hypoxia and by oncogene signaling. In addition, VEGF ligands can be sequestered in the extracellular matrix in latent forms that are subject to release and activation by extracellular matrix-degrading proteases (e.g., MMP-9). FGFs sustain tumor angiogenesis when their expression is chronically upregulated. On the other hand, TSP-1 binds transmembrane receptors displayed by endothelial cells and thereby evokes suppressive signals that can counteract proangiogenic stimuli.

Angiogenesis was previously thought to be important only when rapidly growing macroscopic tumors had formed, but recent data indicate that angiogenesis is induced surprisingly early during the multistage development of invasive cancers, contributing to the microscopic premalignant phase of neoplastic progression. However, the blood vessels within tumors are typically aberrant: precocious capillary sprouting, convoluted and excessive vessel branching, distorted and enlarged vessels, erratic blood flow, microhemorrhaging, leakiness, and abnormal levels of endothelial cell proliferation and apoptosis.

8.1.6 Activating Invasion and Metastasis

Local invasion and distant metastasis are salient characteristics of cancer cells, but their underlying mechanisms are not well understood. During invasion and metastasis, cancer cells typically develop alterations in their shape, as well as in their attachment to other cells and to the extracellular matrix. These alterations are largely due to the loss of E-cadherin, which is a cell-to-cell adhesion molecule that helps to assemble epithelial cell sheets and maintain the quiescence of the cells within these sheets. Increased (decreased) expression of E-cadherin antagonizes (potentiates) invasion and metastasis. N-cadherin is another adhesion molecule, which is normally expressed in migrating neurons and mesenchymal cells during organogenesis. It is found upregulated in many invasive carcinoma cells. Nevertheless, the master regulators of invasion and metastasis are largely unknown or undetermined.

The invasion-metastasis cascade consists of a succession of changes. Cancer cells initiate local invasion and intravasate into nearby blood and lymphatic vessels, whereby they can be transmitted through the lymphatic and hematogenous systems. Extravasation then ensues, whereby cancer cells escape from the lumina of blood and lymphatic vessels into the parenchyma of distant tissues, leading to the formation of micrometastases, small nodules of cancer cells. Finally, micrometastatic lesions grow into macroscopic tumors (colonization).

8.1.7 Reprogramming Energy Metabolism

The uncontrolled cell growth and proliferation of cancer cells must be supported by greatly increased energy metabolism. In cancer cells, the pattern of energy metabolism is altered, as first discovered by Otto Warburg: even in the presence of oxygen, cancer cells prefer to use glycolysis for energy production, a state that has been termed "aerobic glycolysis" or the "Warburg effect." Given that glycolysis is about 18-fold less efficient than oxidative phosphorylation in ATP production, such a reprogramming is highly counterintuitive. Indeed, the energy demand is already high, why would cancer cells use an inefficient method to generate ATP and waste a large quantity of precious glucose unnecessarily? A reasonable explanation is as follows. Rapid cell growth requires not only energy, but also a large quantity of substrates (amino acids, nucleosides, etc.) for biosynthesis of macromolecules and organelles. And these substrates can be extracted from glucose during glycolysis, which diverts glycolytic intermediates into various biosynthetic pathways (e.g., the pentose phosphate pathway). The explanation may also be applicable to rapidly dividing embryonic tissues, which seem to manifest the Warburg effect as well.

To enable the Warburg effect, a continuous influx of a large quantity of glucose is probably necessary. Cancer cells can achieve this through upregulating GLUTs and driving them to work incessantly. It is well known that activated AKT is a master regulator of GLUT expression or translocation to the plasma membrane. A constitutive activation of AKT would enable continuous glucose uptake and the subsequent Warburg effect. In the next chapter, the underlying mechanism of a sustained AKT activation (even with a complete withdrawal of growth factors) will be revealed by mathematically analyzing the PI3K-AKT-TOR pathway. There are other oncoproteins that upregulate glucose transporters and multiple enzymes of the glycolytic pathway, such as hypoxia-inducible factors (HIFs), RAS, and MYC. These genetic changes not only confer the hallmark capabilities of cell proliferation, but also allow cancer cells to avoid cytostatic controls and to attenuate apoptosis.

8.1.8 Evading Immune Destruction

Immune surveillance theory proposes that the immune system constantly monitors cells and tissues and that such immune surveillance is responsible for recognizing and eliminating the vast majority of incipient cancer cells and thus nascent tumors. According to this theory, solid tumors must have developed mechanisms to avoid detection by the various arms of the immune system and to limit the extent of immunological killing.

Immune surveillance theory is supported by the fact that immunocompromised individuals have a dramatic increase of risk factors for certain cancers. Although most of these cancers are induced by viruses, some of them are not virus related. In certain experimental models, it was observed that tumors arose more frequently and grew more rapidly in immunodeficient mice relative to immunocompetent controls. Deficiencies in $CD8^+$ cytotoxic T cells, $CD4^+$ helper T cells, natural killer (NK) cells, each or in combination, led to demonstrable increases in tumor incidence. Therefore, both the innate and adaptive cellular arms of the immune system participate in immune surveillance and tumor eradication.

8.2 OBESITY

Obesity is a state in which excess fat has accumulated in the body to the extent that it may have an adverse effect on health, leading to increased health problems. Body fat composition can be estimated by body mass index (BMI), which is defined as the individual's body mass (in kilograms) divided by the square of his or her height (in meters). A BMI of 18.5 to 25 may indicate optimal weight; a BMI lower than 18.5 is considered underweight; a BMI above 25, 30, and 40 is considered as overweight, obese, and morbidly obese, respectively. Obesity can be further evaluated in terms of fat distribution via the waist–hip ratio.

Obesity is viewed as one of the most serious public health problems of the twenty-first century. From 1971 to 2000, obesity rates in the United States increased from 14.5% to 30.9%. Obesity increases the likelihood of many diseases, such as cardiovascular disease, type 2 diabetes, obstructive sleep apnea, certain types of cancer, osteoarthritis, and asthma. In particular, obesity is closely related to type 2 diabetes. The increase of body fat alters the body's response to insulin, leading to insulin resistance and a prediabetic state. Although some obese people eat little yet gain weight due to a slowing down of metabolism, an average obese person has a greater energy expenditure than his thin counterparts due to the energy required to maintain an increased body mass.

Obesity's fundamental cause is long-term energy imbalance — ingesting more calories than one spends over weeks, months, and years. Of the many contributors to energy imbalance today, plentiful sugar may be the worst. The close relationship between obesity and excessive food intake is demonstrated by an increase of food energy consumed that coincides with the period from 1971 to 2000, during which obesity rates in the United States increased from 14.5% to 30.9%. Most of this extra food energy did not come from fat consumption, but from an increase in carbohydrate consumption (sweetened beverages, potato chips, etc.), which now accounts for almost 25% of daily food energy in young adults in America.

A sedentary lifestyle also plays a significant role in obesity. Worldwide there has been a large shift toward less physically demanding work and currently at least 60% of the world's population gets insufficient exercise. Also, the World Health Organization indicates that people worldwide are taking up less physically active recreational pursuits. In contrast, physically inactive recreations increase, such as television watching and computer games. The emergence of Wii, however, appears to be a good thing, because it necessitates physical activities during the game.

Obesity and diabetes clearly have a genetic basis. They are very harmful, yet they are quite common. What makes it even more surprising is that they became common only *recently*. Polymorphism in various genes controlling appetite and metabolism predispose to obesity when sufficient food energy is present. For example, people with two copies of the FTO gene (fat mass and obesity associated gene) have been found on average to weigh 3–4 kg more and have a 1.67-fold greater risk of obesity compared with those without the risk allele. In people with early-onset severe obesity (defined by an onset before 10 years of age and a BMI over three standard deviations above normal), 7% harbor a single point DNA mutation. Studies focusing on inheritance patterns (rather than on specific genes) have found that 80% of the offspring of two obese parents were also obese, in contrast to less than 10% of the offspring of two parents who were of normal weight.

8.3 DIABETES MELLITUS

Diabetes mellitus, or diabetes for short, is a group of metabolic diseases in which a person has high blood sugar, either because the pancreas does not produce enough insulin, or because cells do not respond to the insulin that is produced. This high blood sugar produces the classic symptoms of polyuria (frequent urination), polydipsia (increased thirst), and polyphagia (increased hunger). There are three main types of diabetes. The epidemic of diabetes has a huge associated cost in terms of health care dollars as well as human morbidity and mortality.

Type 1 diabetes results from the body's failure to produce insulin and requires the person to inject insulin or wear an insulin pump. This form was previously referred to as "insulin-dependent diabetes mellitus" (IDDM) or "juvenile diabetes."

Type 2 diabetes results from insulin resistance, a condition in which cells fail to use insulin properly. Progression into the late stage of this type of diabetes often leads to the failure of the pancreas and insulin deficiency as a consequence. This form was previously referred to as "non-insulin-dependent diabetes mellitus" (NIDDM) or "adult-onset diabetes." In recent years, children are increasingly plagued by obesity and the prevalence of "adult-onset diabetes" in children is approaching that of juvenile diabetes.

Gestational diabetes is a condition in which women without previously diagnosed diabetes exhibit high blood glucose levels during pregnancy (especially during the third trimester). In general, babies born to mothers with untreated gestational diabetes are typically at increased risk of problems such as being overly large for gestational age. This might be due to the high glucose concentration in the mother's blood that provides excessive nutrients to the fetus. Gestational diabetes affects 3% to 10% of pregnancies, depending on the population studied. This high incidence rate may imply that gestational diabetes is a natural phenomenon to some degree. It is very likely that gestational diabetes is an optimal, useful mechanism to shunt the limited glucose from the mother's muscles and divert it to the fetus, provided that the mother is suffering from undernutrition. Because the undernutritional condition is largely removed in many modern human populations, this protective mechanism may cause detrimental effects for certain people (see Chapter 13).

All forms of diabetes have been treatable since insulin became available in 1921, and type 2 diabetes may be controlled with medications. Both types 1 and 2 are chronic conditions that cannot be cured. Pancreas transplants have been tried with limited success in type 1 diabetes; gastric bypass surgery has been successful in many with morbid obesity and type 2 diabetes. Mild gestational diabetes usually resolves after delivery. Diabetes without proper treatment can cause many complications. Acute complications include hypoglycemia, diabetic ketoacidosis, or nonketotic hyperosmolar coma. Serious long-term complications include cardiovascular disease, chronic renal failure, and diabetic retinopathy (retinal damage). Adequate treatment of diabetes is thus important, as well as blood pressure control and lifestyle factors such as smoking cessation and maintaining a healthy body weight.

As of 2012, it is estimated that 346 million people all over the world have type 2 diabetes. Its incidence increases rapidly, and by 2030, it is estimated that this number will almost double. The epidemic of type 2 diabetes is closely related to that of obesity, for which reason "the twin epidemics" is often used to refer to obesity and type 2 diabetes. Several hypotheses (see below) have been proposed to explain the escalation of the twin epidemics.

8.3.1 Thrifty Gene Hypothesis

In 1962, geneticist James V. Neel proposed the thrifty gene hypothesis [23] to resolve a fundamental problem: how obesity and diabetes, diseases with an apparent genetic basis and with such negative effects, may have been favored by the process of natural selection. Thrifty gene refers to genes that enable individuals to efficiently store food energy as fat during periods while foods are available. Because the stored fat can be utilized during times of famine, which occurs frequently, these genes were certainly advantageous in the course of evolution. In the modern setting of sedentary lifestyles and unrestricted access to high-caloric foods, thrifty genes have been suggested to underlie the twin epidemics of obesity and diabetes. Because famine never comes, fats accumulate year after year, leading to widespread chronic obesity and related health problems like diabetes. To explain insulin resistance that is associated with type 2 diabetes, Neel suggested that evolutionary pressure to preserve glucose for use by the brain during starvation led to a genetic propensity toward insulin resistance in peripheral tissues.

In a 1998 review, Neel extended the original hypothesis to a more complex theory of several related diseases such as diabetes, obesity, and hypertension being caused by physiological systems adapted for an older environment being pushed beyond their limits by environmental changes. Thus, one possible remedy for these diseases is changing diet and exercise activity to more closely reflect that of the ancestral environment.

Type 2 diabetes and obesity have a clear genetic component but are rarely monogenic. There are likely to be multiple thrifty genes, and the inheritance of several polymorphisms leading to small differences in expression can make populations more or less susceptible to obesity and diabetes. One thrifty gene candidate is the gene that encodes peroxisome proliferator-activated receptor (PPAR-γ), a transcription factor that is the master regulator of adipogenesis (the differentiation of preadipocytes into adipocytes) and lipid accumulation in adipocytes. Another thrifty gene candidate is the gene that encodes leptin, a hormone produced by adipose tissue, the absence of which leads to obesity and insulin resistance in rodents and humans. Leptin functions physiologically as a signal of energy stores, inhibiting food intake and accelerating energy metabolism. During starvation, it is the fall in circulating leptin levels that triggers increased appetite and decreased metabolic rate. Because a reduction in leptin levels appears to be the physiological signal for a thrifty metabolic response, leptin itself must have been evolutionarily selected for another function. Indeed, leptin favors survival of the species by conferring a reproductive advantage to individuals who are nutritionally fit.

8.3.2 Thrifty Phenotype Hypothesis

C. Nicholas Hales and David J. P. Barker noticed the epidemiological associations between poor fetal and infant growth and the subsequent development of type 2 diabetes. In 1992, they proposed the thrifty phenotype hypothesis on the basis of clinical observations that poor fetal and/or postnatal nutrition is associated with obesity and type 2 diabetes later in life [14]. Instead of the "thrifty factors" arising from genetics, the thrifty phenotype hypothesis proposes that environmental factors (e.g., nutrients and their interaction with genes) are the dominant cause of type 2 diabetes. This hypothesis posits that fetal malnutrition alters metabolic pathways that result in tissue adaptations favoring the thrifty use of nutrients in utero and in postnatal life, thereby leading to obesity and diabetes in the setting of subsequent adequate nutrition. That is, poor fetal and infant growth can produce permanent changes in glucose-insulin metabolism, which include reduced capacity for insulin secretion (poor development of pancreatic beta cell mass and function) and insulin resistance.

The thrifty phenotype hypothesis emphasizes a memory in the later life of the fetal/neonatal environment. Such a memory may be conferred by epigenetic regulation of gene expression, which involves chemical modification of chromatin by enzymes such as sirtuins, whose activities are linked to cellular energy stores and, in lower organisms, interface with insulin signaling pathways.

8.3.3 Drifty Gene Hypothesis

British biologist John Speakman proposed the drifty gene hypothesis as an alternative to the thrifty gene hypothesis [29]. The main idea was that genes favoring obesity have not been positively selected in our past, but rather have been subject to random drift because of an absence of selection (selectively neutral). As by definition genes cannot be under selection and also drifting with no selection, the drifty gene and thrifty gene ideas are mutually incompatible.

One example of a drifty gene was given in the context of the "predation release hypothesis." The hypothesis argues that our body fatness was historically regulated by a system that involves upper and lower intervention limits, as appears to be the case in some wild animals today. The lower intervention limit is set by the risk of starvation, and the upper intervention limit is set by the risk of predation. Two million years ago, humans went through a transition in the exposure to predation risk because at that time humans developed social behavior, weapons, and fire. This effectively removed any selection maintaining the position of the upper intervention point. Over the past two million years, the genes defining the upper intervention point have been subject to random mutation and drift. Consequently, when embedded in modern societies where energy is freely available, individuals move to their drifted upper intervention points. This may generate a pattern of susceptibility to obesity that reasonably mimics the form of the modern epidemic.

8.3.4 Mechanistic Link Between Obesity and Diabetes

The above hypotheses are of great interest but are inherently speculative and difficult to prove. Mechanistic studies are thus important to reveal the obesity-diabetes connections, which involve numerous fat-derived molecules and inflammation.

Adipose tissue is both a passive fuel depot and an active endocrine organ that communicates with the brain and peripheral tissues by secreting hormones regulating appetite and metabolism. Several factors affect the functions of adipose tissue: the location (visceral versus subcutaneous), the size of adipocytes, and glucose metabolism in adipocytes. Adipose tissue secretes a large number of molecules to modulate glucose metabolism and insulin action, including leptin, adiponectin, resistin, tumor necrosis factor-alpha (TNF-α), interleukin-6 (IL-6), monocyte chemoattractant protein-1 (MCP-1), and of course, free fatty acids (FFAs). Among these molecules, adiponectin is special in that it enhances insulin action yet circulates at reduced levels in obesity, whereas the other molecules are generally elevated in individuals with obesity or type 2 diabetes. While leptin and adiponectin enhance insulin responsiveness, resistin, TNF-α, IL-6, MCP-1, and FFAs contribute to insulin resistance in liver and muscle in obesity.

The close relationship between obesity and inflammation is suggested by the fact that, besides adipocyte, adipose tissue also contains macrophages. Indeed, it has become evident that macrophage infiltration of adipose tissue is characteristic of obesity. Both adipocytes and macrophages express PPAR-γ, the master regulator that promotes adipogenesis and lipid storage in adipocytes. Some proteins secreted by adipose tissue are actually produced by macrophages, suggesting their roles in innate immunity, a relatively primitive defense mechanism against infection. Cytokines such as TNF-α and IL-6 are produced by macrophages as well as by adipocytes; they act directly on inflammatory cells and also contribute indirectly to inflammation by acting on the liver to produce acute-phase proteins. These cytokines also induce suppressor of cytokine signaling-3 (SOCS-3), an intracellular signaling molecule elevated in obesity that impairs both leptin and insulin signaling. The anatomic blurring of the line between adipocytes and macrophages is paralleled by the tissue expression of resistin, which is expressed exclusively in adipocytes in mice but predominantly in macrophages in humans. It is thus possible that the commonalities of adipocyte and macrophage function are remnants of an ancestral evolutionary adaptation.

The above discussion implies that obesity is an inflammatory state that induces insulin resistance, the hallmark of prediabetes. Because inflammation is a natural response to pathogen infection, this might imply mild insulin resistance might be beneficial in an inflammatory response to clear pathogens. Perhaps the response to infection is more effective when glucose is shunted from muscles (peripheral insulin resistance) to the inflammatory cells involved in the immune response and tissue repair. Peripheral insulin resistance can be induced by inflammatory factors secreted from macrophages and adipocytes, which interfere with the interaction between IRS and PI3K and cause an enlarged threshold of insulin responsiveness.

In the following chapters, mathematical analysis of the PI3K-AKT-TOR pathway will reveal the network-level control mechanism of insulin resistance. It will also be explained how obesity can induce diabetes.

9 Mathematical Modeling of the PI3K-AKT-TOR Pathway

9.1 BACKBONE NETWORK STRUCTURE

In Chapter 7, the structure of the PI3K-AKT-TOR pathway (illustrated in Figure 7.1.) was explained in detail. Given the complexity of biochemical reactions and information processing in the network, mathematical modeling is needed to capture the essentials of the pathway and to reveal its control mechanisms, which may ultimately lead to the understanding of the normal functioning of the pathway and how its malfunction can lead to various pathologic states.

The key to mathematical modeling is simplification without losing essentials. Therefore, the complex network in Figure 7.1 is simplified into a backbone network structure as illustrated in Figure 9.1. The simplification is according to the following considerations. First, because AKT is the master regulator of the pathway, the detailed phosphorylation and dephosphorylation of AKT are largely retained in the backbone structure. However, we do not consider separately the two phosphorylation sites (T308 and S473) and their respective protein kinases (PDK1 and TORC2). Instead, we group PDK1 and TORC2 as a single enzyme E_1 catalyzing the phosphorylation (activation) of AKT. As such, the two phosphatases are grouped into a single enzyme E_2 catalyzing the dephosphorylation (deactivation) of AKT. Second, molecules along a linear cascade are represented by only three molecules: two ending ones and a middle one. For example, the cascade *pAKT* ⊣ *TSC1,2* ⊣ *Rheb-GTP* → *TOR* → *S6K* ⊣ *pIRS* is simplified to *pAKT* → *TOR* ⊣ *pIRS*. Note that the signs of the simplified structure follow the rules of multiplication, with → acting as a positive number and ⊣ acting as a negative number. Therefore, *pAKT* ⊣ *TSC1,2* ⊣ *Rheb-GTP* → *TOR* is simplified to *pAKT* → *TOR*, because two negatives make one positive; *TOR* → *S6K* ⊣ *pIRS* is simplified to *TOR* ⊣ *pIRS*, because one negative and one positive make one negative. Here TOR actually represents TORC1. These simplifications lead to a backbone network structure, as illustrated in Figure 9.1. The backbone structure can be divided into three components (colored in green, blue, and red) according to their functions: the input component (colored in green), the output component (colored in blue), and the feedback component (colored in red). In the following insulin is used to elucidate the backbone network structure, although the network responds to many other growth factors as well.

9.1.1 INPUT COMPONENT

Insulin binds with the insulin receptor (IR) and stimulates it. The insulin receptor thus has an activity level, called the insulin level and denoted by I. In the following, the

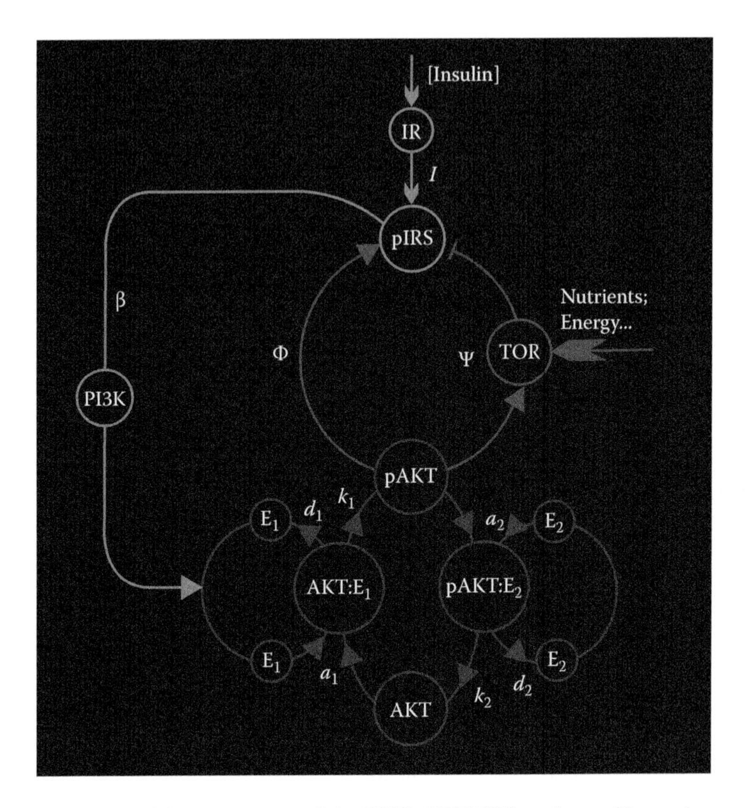

FIGURE 9.1 The backbone structure of the PI3K-AKT-TOR pathway. The pathway can be divided into three components: the input (colored in green), the output (colored in blue), and the feedback (colored in red). See color insert.

description of the pathway will always start from I. To relate the insulin level I to the actual insulin concentration, one can use

$$I = \frac{[\text{Insulin}]}{\varepsilon_1 + \varepsilon_2[\text{Insulin}]}, \tag{9.1}$$

where ε_1, ε_2 are two parameters. The insulin level I stimulates IRS and leads to the formation of phospho-IRS-tyrosine (pIRS). pIRS then stimulates PI3K, which in turn activates the enzyme group E_1. The gain of the input component is denoted by β.

9.1.2 Output Component (Blue)

The output component is the core of the pathway. It is the phosphorylation and dephosphorylation cycle (PdPC) of AKT $\rightleftharpoons$ pAKT, where pAKT stands for phospho-AKT (i.e., the activated AKT). The cycle is mediated by two opposing groups of enzymes: E_1 and E_2, which form a recurring network motif, known as the Goldbeter–Koshland loop. The biochemical reactions of this PdPC can be described by

$$\text{AKT} + \text{E}_1 \underset{d_1}{\overset{a_1}{\rightleftharpoons}} \text{AKT} : \text{E}_1 \xrightarrow{k_1} \text{pAKT} + \text{E}_1, \tag{9.2}$$

$$\text{pAKT} + \text{E}_2 \underset{d_2}{\overset{a_2}{\rightleftharpoons}} \text{pAKT} : \text{E}_2 \xrightarrow{k_2} \text{AKT} + \text{E}_2. \tag{9.3}$$

The Goldbeter–Koshland kinetics are essentially a pair of Michaelis–Menten kinetics, which are described in Section 5.4.

9.1.3 Feedback Component (Red)

pAKT regulates pIRS both positively and negatively. On one hand, pAKT directly interacts with pIRS and enhances its activity (pAKT $\rightarrow$ pIRS), which generates a positive feedback loop. On the other hand, pAKT reduces pIRS activity indirectly (pAKT $\rightarrow$ TOR $\dashv$ pIRS), which forms a negative feedback loop. The strengths of positive and negative feedbacks are denoted by Φ and Ψ, respectively. Note that Ψ reflects TOR activity, which is regulated by many factors, such as the intracellular nutrient level, energy status, and so on.

9.2 MATHEMATICAL MODEL

Based on the backbone structure (Figure 9.1), a mathematical model is developed to describe the biochemical reactions in the network:

$$\frac{d[\text{AKT}]}{dt} = -a_1[\text{AKT}][\text{E}_1] + d_1[\text{AKT} : \text{E}_1] + k_2[\text{pAKT} : \text{E}_2], \tag{9.4}$$

$$\frac{d[\text{AKT} : \text{E}_1]}{dt} = a_1[\text{AKT}][\text{E}_1] - (d_1 + k_1)[\text{AKT} : \text{E}_1], \tag{9.5}$$

$$\frac{d[\text{pAKT}]}{dt} = -a_2[\text{pAKT}][\text{E}_2] + \text{d}_2[\text{pAKT} : \text{E}_2] + \text{k}_1[\text{AKT} : \text{E}_1], \tag{9.6}$$

$$\frac{d[\text{pAKT} : \text{E}_2]}{dt} = a_2[\text{pAKT}][\text{E}_2] - (d_2 + k_2)[\text{pAKT} : \text{E}_2], \tag{9.7}$$

$$\frac{d[\text{pIRS}]}{dt} = I + (\Phi - \Psi[\text{pIRS}])[\text{pAKT}] - \delta[\text{pIRS}], \tag{9.8}$$

$$A_{\text{max}} = [\text{AKT}] + [\text{pAKT}], \tag{9.9}$$

$$E_{1T} = [\text{E}_1] + [\text{AKT} : \text{E}_1], \tag{9.10}$$

$$E_{2T} = [\text{E}_2] + [\text{pAKT} : \text{E}_2], \tag{9.11}$$

$$E_{1T} = \beta[\text{pIRS}], \tag{9.12}$$

where

a_1, d_1, and k_1 = the rate constants of association, dissociation, and catalyzation of the reaction AKT $\rightarrow$ pAKT.

a_2, d_2, and k_2 = the rate constants of association, dissociation, and catalyzation of the reaction pAKT $\rightarrow$ AKT.

β = the strength of the input component.

I = the insulin level.

δ = the decay rate of pIRS.

$$\Phi = \text{the gain of the positive feedback loop.}$$
$$\Psi = \text{the gain of the negative feedback loop.}$$
$$A_{\max} = \text{the concentration of total AKT, namely, } A_{\max} = [\text{AKT}]_{\text{tot}}.$$
$$E_{1T} = \text{the concentration of total } \text{E}_1.$$
$$E_{2T} = \text{the concentration of total } \text{E}_2.$$
$$K_{m1} = (d_1 + k_1)/a_1 = \text{the Michaelis constant of the reaction AKT} \rightarrow \text{pAKT}.$$
$$K_{m2} = (d_2 + k_2)/a_2 = \text{the Michaelis constant of the reaction pAKT} \rightarrow \text{AKT}.$$

The parameters K_{m1} and K_{m2} are not present in Equations (9.4–9.12) per se. They are listed here because they will soon play an important role. To be stricter, one should use

$$A_{\max} = [\text{AKT}] + [\text{pAKT}] + [\text{AKT} : \text{E}_1] + [\text{pAKT} : \text{E}_2] \tag{9.13}$$

instead of Equation (9.9). Here we assume that the total AKT is in much excess over E_1 and E_2 so that the compound concentrations [AKT:E_1] and [pAKT:E_2] can be neglected in Equation (9.13). This assumption is generic and was widely used in studying the Goldbeter–Koshland kinetics [11].

Although there are five state variables ([AKT], [AKT:E_1], [pAKT], [pAKT:E_2], [pIRS]) in the model, only four of them are independent. The model is thus a four-dimensional nonlinear dynamical system. Equations (9.4–9.7) describe the Goldbeter–Koshland kinetics AKT $\rightleftharpoons$ pAKT (which was illustrated by Equations (9.2) and (9.3)). Equations (9.9–9.11) describe the conservation of substances during the chemical reactions. Equation (9.12) makes the approximation that E_{1T} is proportional to [pIRS], with the gain β. Equation (9.8) describes the kinetics of pIRS. Of note, the negative feedback $-\Psi$[pIRS][pAKT] is proportional to both [pIRS] and [pAKT], due to the law of mass action. In contrast, the positive feedback Φ[pAKT] is proportional to [pAKT] only.

9.3 MASTER EQUATION

The PI3K-AKT-TOR pathway has been modeled by Equations (9.4–9.12), which has five state variables ([AKT], [AKT:E_1], [pAKT], [pAKT:E_2], and [pIRS]) and many parameters ($a_1, d_1, k_1, \beta, I, \ldots$).

With the initial values (i.e., values at time $t = 0$) of the state variables given, the temporal changes of the state variables can be obtained by integrating Equations (9.4–9.12). In biology, such a dynamical change usually settles into a steady state or a limited cycle, as explained in Chapter 6. Because a limited cycle does not serve any known biological function of the PI3K-AKT-TOR pathway, in this book we assume the dynamical change finally stabilizes at a steady state.

At the steady state, the concentrations of molecules do not change over time. That is, [X] = constant and $d[\text{X}]/dt = 0$, where [X] represents the concentration of molecule X. Therefore, the steady states of the PI3K-AKT-TOR pathway can be determined by making the left-hand side of Equations (9.4–9.8) zero, which results in the following set of algebraic equations:

$$-a_1[\text{AKT}][\text{E}_1] + d_1[\text{AKT: E}_1] + k_2[\text{pAKT} : \text{E}_2] = 0, \tag{9.14}$$
$$a_1[\text{AKT}][\text{E}_1] - (d_1 + k_1)[\text{AKT: E}_1] = 0, \tag{9.15}$$

$$-a_2[\mathrm{pAKT}][\mathrm{E}_2] + d_2[\mathrm{pAKT{:}\ E}_2] + k_1[\mathrm{AKT : E}_1] = 0, \tag{9.16}$$
$$a_2[\mathrm{pAKT}][\mathrm{E}_2] - (d_2 + k_2)\,[\mathrm{pAKT{:}\ E}_2] = 0, \tag{9.17}$$
$$I + (\Phi - \Psi[\mathrm{pIRS}])\,[\mathrm{pAKT}] - \delta[\mathrm{pIRS}] = 0, \tag{9.18}$$
$$[\mathrm{AKT}] + [\mathrm{pAKT}] = A_{\max}, \tag{9.19}$$
$$[\mathrm{E}_1] + [\mathrm{AKT{:}\ E}_1] = E_{1T}, \tag{9.20}$$
$$[\mathrm{E}_2] + [\mathrm{pAKT{:}\ E}_2] = E_{2T}, \tag{9.21}$$
$$\beta[\mathrm{pIRS}] = E_{1T}. \tag{9.22}$$

After some maneuvers (see Appendix B), the above equations reduce to an algebraic equation of the third degree,

$$G(I, A, \mathbf{p}) = 0, \tag{9.23}$$

called *the master equation of AKT activation.* Here $G\,(I, A, \mathbf{p})$ abbreviates a long expression:

$$G = \theta A^3 + \big((K_1 + K_2)\phi - (K_1 + 1)\theta + \beta\gamma I - 1\big)\,A^2$$
$$+\big(K_1 + 1 + (K_2 - 1)\beta\gamma I - K_2\phi\big)\,A - K_2\beta\gamma I. \tag{9.24}$$

In the master equation,

$$\mathbf{p} = (\phi, \theta, \beta, \gamma, K_1, K_2) \tag{9.25}$$

is a vector of parameters, where

$$\gamma = \frac{k_1}{\delta k_2 E_{2T}},$$
$$\phi = \Phi\beta\gamma A_{\max},$$
$$\theta = (\Phi\beta\gamma - \Psi/\delta)\,A_{\max},$$
$$K_1 = \frac{K_{m1}}{A_{\max}} = \frac{d_1 + k_1}{a_1 A_{\max}},$$
$$K_2 = \frac{K_{m2}}{A_{\max}} = \frac{d_2 + k_2}{a_2 A_{\max}}.$$

Except for β, which has appeared in Equations (9.4–9.22), the other parameters in $\mathbf{p}$ show up here for the first time. Each of the new parameters is a combination of several original parameters. For example, γ is a combination of k_1, k_2, δ, and E_{2T}.

In the master equation, the insulin level I is a distinguished parameter, because it triggers activation. Being special, I is not included in the vector $\mathbf{p}$. It can be regarded as the input to the pathway. In the master equation, the symbol A represents the percentage of the activated AKT in the steady state, i.e.,

$$A = [\mathrm{pAKT}]/A_{\max}.$$

It can be regarded as the output from the PI3K-AKT-TOR pathway. The quantitative relationship between I and A, namely, the stimulus-response curve, will play a pivotal role in our theory.

9.4 SIMPLIFIED MASTER EQUATION

In the mathematical model, the positive feedback Φ[pAKT] and the negative feedback $-\Psi$[pIRS][pAKT] are not exactly in opposite directions, due to the additional factor [pIRS] in the negative feedback term. In this section, the combined action of positive and negative feedback is approximated by a single feedback $(\Phi - \epsilon\Psi)$[pAKT], where ϵ is a constant that can be considered as the averaged [pIRS], in some sense. It will turn out that the approximation is valid in biologically relevant settings (see Appendix E). By this approximation, the steady-state model becomes

$$-a_1[\mathrm{AKT}][\mathrm{E}_1] + d_1[\mathrm{AKT{:}\ E}_1] + k_2[\mathrm{pAKT : E}_2] = 0, \tag{9.26}$$

$$a_1[\mathrm{AKT}][\mathrm{E}_1] - (d_1 + k_1)[\mathrm{AKT{:}\ E}_1] = 0, \tag{9.27}$$

$$-a_2[\mathrm{pAKT}][\mathrm{E}_2] + d_2[\mathrm{pAKT{:}\ E}_2] + k_1[\mathrm{AKT : E}_1] = 0, \tag{9.28}$$

$$a_2[\mathrm{pAKT}][\mathrm{E}_2] - (d_2 + k_2)[\mathrm{pAKT{:}\ E}_2] = 0, \tag{9.29}$$

$$I + (\Phi - \epsilon\Psi)[\mathrm{pAKT}] - \delta[\mathrm{pIRS}] = 0, \tag{9.30}$$

$$[\mathrm{AKT}] + [\mathrm{pAKT}] = A_{\max}, \tag{9.31}$$

$$[\mathrm{E}_1] + [\mathrm{AKT{:}\ E}_1] = E_{1T}, \tag{9.32}$$

$$[\mathrm{E}_2] + [\mathrm{pAKT{:}\ E}_2] = E_{2T}, \tag{9.33}$$

$$\beta[\mathrm{pIRS}] = E_{1T}. \tag{9.34}$$

The above equations can be reduced to a third-degree algebraic equation, called *the simplified master equation*. In the following, the simplified master equation is presented in two forms, which are essentially the same. They will be used in different contexts, just for the sake of convenience.

9.4.1 Form I

$$G(I, A, \mathbf{p}) = 0, \tag{9.35}$$

where G abbreviates an expression slightly different from Equation (9.24):

$$\begin{aligned} G = {} & \alpha\beta A^3 + \big((K_2 - 1)\alpha\beta + \beta\gamma I - 1\big)A^2 \\ & + \big(K_1 + 1 + (K_2 - 1)\beta\gamma I - K_2\alpha\beta\big)A - K_2\beta\gamma I. \end{aligned} \tag{9.36}$$

The vector

$$\mathbf{p} = (\alpha, \beta, \gamma, K_1, K_2)$$

is also different from Equation (9.25). It has a new parameter

$$\alpha = (\Phi - \epsilon\Psi)\gamma A_{\max}$$

to replace ϕ and θ in Equation (9.25).

9.4.2 Form II

$$\tilde{G}(\tilde{I}, A, \tilde{\mathbf{p}}) = 0, \tag{9.37}$$

where

$$\tilde{G} = \tilde{\theta}A^3 + \big((K_2 - 1)\tilde{\theta} + \tilde{I} - 1\big)A^2 \\ +\big(K_1 + 1 + (K_2 - 1)\tilde{I} - K_2\tilde{\theta}\big)A - K_2\tilde{I}, \quad (9.38)$$

and

$$\tilde{\mathbf{p}} = (\tilde{\theta}, K_1, K_2), \\ \tilde{\theta} = \alpha\beta, \\ \tilde{I} = \beta\gamma I.$$

Note the two new parameters $\tilde{\theta}$ and $\tilde{I}$. The parameter $\tilde{\theta}$ is used to replace α and β. The parameter $\tilde{I}$, being proportional to the insulin level I, is used here to replace I. Form II of the simplified master equation has only three parameters and appears to be simpler than form I.

9.5 RESPONSE CURVE

The response curve is an important concept in engineering. The normal functioning of an electronic circuit, for example, can be defined by a response curve of certain shapes. Abnormal shapes of the response curve may signal malfunctions of the electronic circuit. As a well-organized circuit of molecular interactions, the PI3K-AKT-TOR pathway has great similarities with an electronic circuit, with its response curve being a crucial indicator of its biological functions.

Then, what is the response curve of the PI3K-AKT-TOR pathway in terms of the mathematical model? The master equation $G(I, A, \mathbf{p}) = 0$ constrains the relationship among I, A, and $\mathbf{p}$, in the steady state. If any two of the three are given, then the third can be determined from the master equation. If only one is given, then the quantitative relationship between the remaining two can be determined as a curve. That is, given a set of $\mathbf{p}$, one can determine a curve $A(I)$. When $\mathbf{p}$ changes, the shape of $A(I)$ also changes. This curve is called the stimulus-response curve, or the response curve for short.

Let the blue curve in Figure 9.2(a) be a response curve. The stability of this curve is explained as follows, by using several thought experiments of perturbations. We first discuss scenarios with I fixed at 1. One sees that the steady state A is 0.5. Suppose an instantaneous fluctuation perturbs A to a new value of 0.3. The system then undergoes a bout of dynamical evolution (all the state variables change continuously over time), until the system state returns to the steady state (with $A = 0.5$). Note that the steady state is the same one as before, because I does not change. This transition is represented by the up arrow in Figure 9.2(a). If an instantaneous fluctuation perturbs A to 0.8, then another dynamical evolution ensues, which finally returns to the same steady state (the down arrow in Figure 9.2(a)).

Let I jump to 2 instantaneously (the dashed arrow in Figure 9.2(b)), while keeping everything else unchanged at that moment. One sees that the new steady state becomes $A = 0.848$. As a consequence, the system has to undergo a round of dynamical evolution, starting from $A = 0.5$, until the new steady state $A = 0.848$ is achieved (the up arrow).

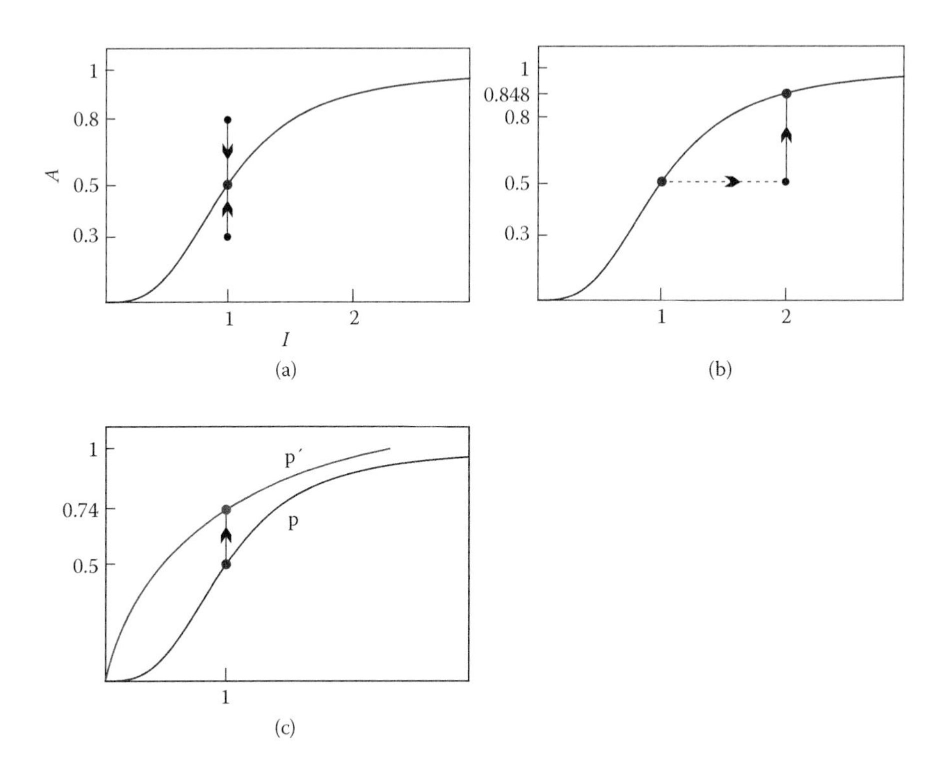

FIGURE 9.2 The response curve and the network state perturbation. (a) Random perturbations of the A value while keeping I fixed. (b) The perturbation of I to 2 causes the network state to change to a new steady state. (c) The change of the parameters makes the entire response curve change. See color insert.

Fix $I = 1$ again but change the parameters from $\mathbf{p}$ to $\mathbf{p}'$. Then the response curve changes from the blue one to the red one in Figure 9.2(c). Because $A = 0.5$ is not the steady state in terms of the parameters $\mathbf{p}'$, a bout of dynamical evolution is activated so that A can finally be stabilized at $A = 0.74$, the new steady state in terms of $\mathbf{p}'$.

9.6 CHARACTERIZATION OF RESPONSE CURVE

The above discussions highlight the importance of characterizing the kinds of response curve of the PI3K-AKT-TOR pathway. The determination of one response curve is easy: one just substitutes a value of $\mathbf{p}$ into Equation (9.35) and then computes the response curve $A(I)$. In geometric terms, a value of $\mathbf{p}$ is a point in the five-dimensional parameter space, and the point corresponds to a specific response curve. By sampling points in the parameter space, one can produce as many response curves as desired. By observing the obtained response curves, one may develop some intuition about the functional output of the pathway.

More importantly, one can classify all the examined points according to the shapes of their corresponding response curves, through which the parameter space can be divided into different regions. Each region includes all the points whose response curves are of the same kind, which is qualitatively different from the kind of response curves of another region. That is, the response curves in the same region are similar to each other (with no qualitative differences). One can then relate these regions to various physiologic or pathologic phenotypes. For example, region 1 may correspond to the normal functioning of the pathway in the resting state, region 2 may correspond to the normal functioning of the pathway during exercise, region 3 may correspond to cancer, region 4 may correspond to some other disease, and so on. In this way, a comprehensive understanding of the pathway may be developed by sampling as many points in the parameter space as possible and then classifying the sampled points.

This approach is, however, time consuming. It is known that computational complexity increases exponentially with the number of parameters. The five-dimensional parameter space is thus a large one. If one samples 10^3 points for one parameter, then the total number of samplings is 10^{15}. Regardless of the search algorithms to be used, one has to visit most of the 10^{15} points to increase the confidence that all the essential kinds of response have been identified. Even if all 10^{15} points have been investigated, one is still not sure how complete the enumeration is, because 10^3 points per parameter may still underestimate the complexity of the PI3K-AKT-TOR pathway. All in all, a great deal of CPU time has to be used for sampling and computation.

The power of mathematics consists in its ability to avoid unnecessary enumerations and to point to the nature of things. In Appendices C, D, and E, I will show how mathematical analysis can help to efficiently divide the parameter space into different regions, with much reduced numerical computations. In Appendix C, the underlying mathematical theory, namely, singularity theory, is introduced. A variety of shapes are also defined: the toggle switch, the irreversible switch, and the monotone type. In Appendix D, singularity theory is applied to the simplified master equation of form II, whereby two boundary surfaces are identified that divide the parameter space into three regions. The results are analytical and essentially require no numerical computations. In Appendix E, singularity theory is applied to the full master equation, which requires few numerical computations. The results are similar to those obtained by the simplified master equation, indicating that the simplified master equation is a valid approximation of the full master equation.

9.6.1 In the Space $\tilde{\mathbf{p}} = (\tilde{\theta}, K_1, K_2)$ (Three Dimensional)

In Appendix D, two boundary surfaces are identified: $\tilde{\theta} = \tilde{\Theta}_1(K_1, K_2)$ (colored in red in Figure 9.3) and $\tilde{\theta} = \tilde{\Theta}_0(K_1, K_2)$ (colored in green in Figure 9.3). The two surfaces divide the space of $\tilde{\mathbf{p}}$ into three regions; each contains all the points whose response curves $A(\tilde{I})$ are qualitatively the same. For the region above the red surface, the corresponding response curves are irreversible switches. For the region between the red surface and the green surface, the corresponding response curves are toggle switches. For the region below the green surface, the corresponding response curves are of monotone type.

Importantly, we prove in Appendix D that other boundary surfaces (which mark the transitions between other kinds of response curve) do not exist. Figure 9.3 thus

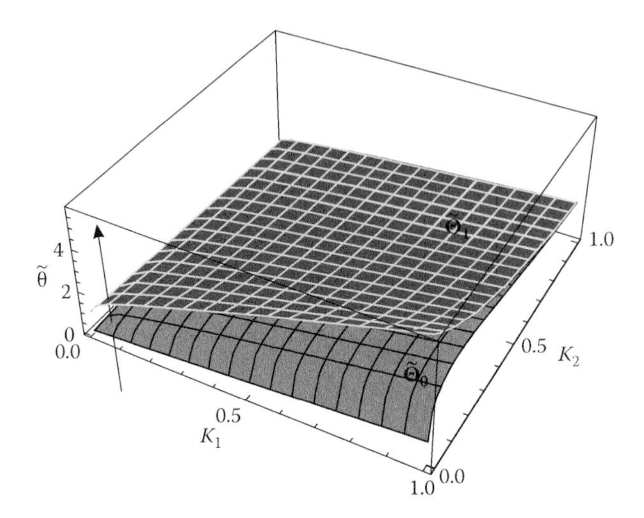

FIGURE 9.3 The space of $\tilde{\mathbf{p}} = \left(\tilde{\theta}, K_1, K_2\right)$. The surface $\tilde{\Theta}_1(K_1, K_2)$ (red) and the surface $\tilde{\Theta}_0(K_1, K_2)$ (green) divide the space into three regions, which from the top down are the irreversible switch, the toggle switch, and the monotone type. See color insert.

provides a complete picture of the steady-state behavior of the PI3K-AKT-TOR pathway.

9.6.2 In the Phase Diagram K Versus $\tilde{\theta}$ (Two Dimensional)

Figure 9.4 presents the diagonal section ($K_1 = K_2 = K$) of Figure 9.3, where the two boundary curves $\tilde{\theta} = \tilde{\Theta}_1(K)$ and $\tilde{\theta} = \tilde{\Theta}_0(K)$ are the diagonal sections of the two surfaces $\tilde{\theta} = \tilde{\Theta}_1(K_1, K_2)$ and $\tilde{\theta} = \tilde{\Theta}_0(K_1, K_2)$, respectively. The two boundaries divide the diagram into three regions, each corresponding to a kind of response curve. In the phase diagram, several points are selected for the presentation of their corresponding response curves $A(\tilde{I})$.

9.6.3 In The $\tilde{\theta}$ Axis (One Dimensional)

In Figure 9.3, the arrow indicates a continuous increase of $\tilde{\theta}$ with fixed $K_1 = K_2 = 0.02$. In Figure 9.5, this $\tilde{\theta}$-axis is reproduced in the horizontal direction, where the dots represent its intersections with the surfaces $\tilde{\theta} = 0$, $\tilde{\theta} = \tilde{\Theta}_0(K_1, K_2)$, and $\tilde{\theta} = \tilde{\Theta}_1(K_1, K_2)$. Response curves of different $\tilde{\theta}$ values are illustrated above the $\tilde{\theta}$-axis.

Monotone type ($\tilde{\theta} < \tilde{\Theta}_0$)**.** The corresponding response curve is a continuous, monotonically increasing function (Figure 9.5(a, b)). As $\tilde{\theta}$ increases, the curve becomes steeper and steeper. When $\tilde{\theta}$ reaches 0, the curve has become switch-like with a large slope (Figure 9.5(b)). As $\tilde{\theta}$ approaches $\tilde{\Theta}_0$ (a small positive number), the slope of the response curve approaches infinity. Note that the value of $\tilde{\Theta}_0$ is

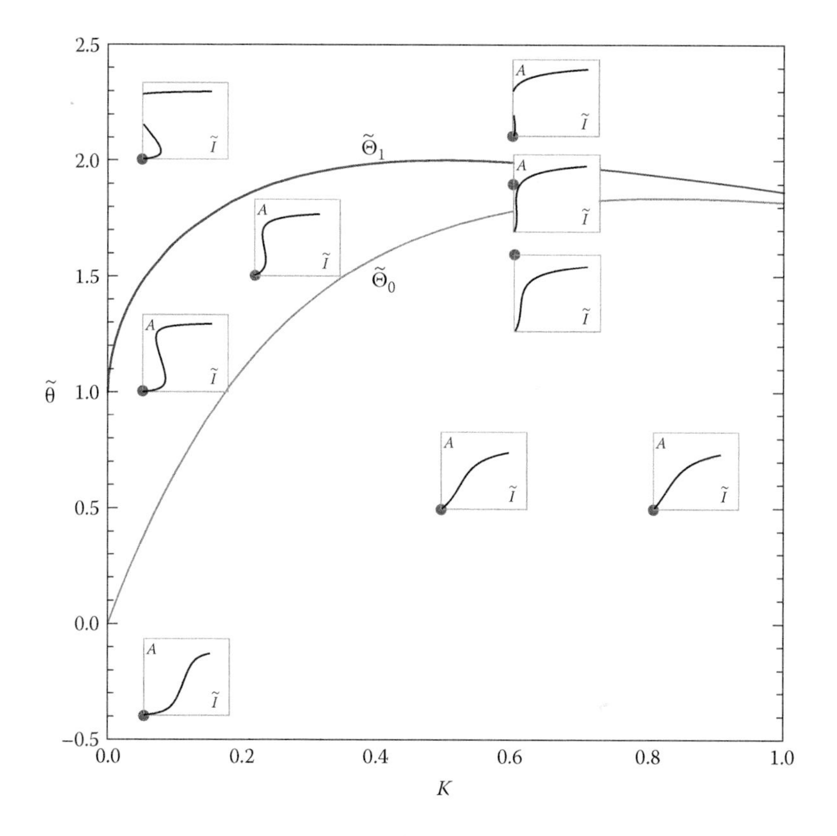

FIGURE 9.4 The phase diagram K versus $\tilde{\theta}$ obtained from the simplified master equation. It is the diagonal section ($K_1 = K_2 = K$) of Figure 9.3. The response curves of some points are shown.

determined by K_1 and K_2: it is 0.15 for the present case $K_1 = K_2 = 0.02$; it is 0.078 for $K_1 = K_2 = 0.01$; and it approaches 0 as K_1 and K_2 approaches 0.

Hysteresis point ($\tilde{\theta} = \tilde{\Theta}_0$). The corresponding response curve has an infinitely large slope, as indicated by the vertical tangent at the switch point $\tilde{I} = \tilde{I}_0$ (Figure 9.5(c)). The response curve is thus a discontinuous function. Therefore, a qualitative transition (from continuous to discontinuous) occurs at $\tilde{\theta} = \tilde{\Theta}_0$. This transition is, however, instantaneous—a further qualitative transition (from a function to a nonfunction) ensues as soon as $\tilde{\theta}$ leaves $\tilde{\Theta}_0$. In brief, $\tilde{\theta} = \tilde{\Theta}_0$ is very special, around which three qualitative entities exist: the continuous function ($\tilde{\theta} < \tilde{\Theta}_0$), the discontinuous function ($\tilde{\theta} = \tilde{\Theta}_0$), and the toggle switch ($\tilde{\theta} > \tilde{\Theta}_0$), which is not a function. The toggle switch (see below) has two different switching points, $\tilde{I}_{\text{off}}$ and $\tilde{I}_{\text{on}}$, indicating that the route of switch-on is different from that of switch-off (hysteresis). This property might be where the name "hysteresis point" comes from.

Toggle switch ($\tilde{\Theta}_0 < \tilde{\theta} < \tilde{\Theta}_1$). The corresponding response curve is *not* a function (Figure 9.5(d)). To be a function, one $\tilde{I}$ value (the preimage) must correspond

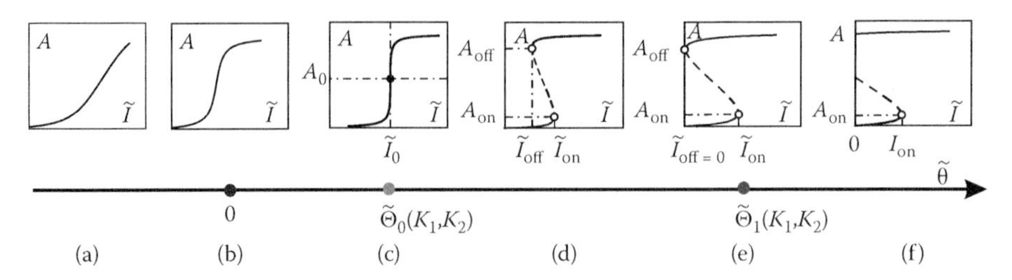

FIGURE 9.5 The deformation of the response curve $A(\tilde{I})$ by the parameter $\tilde{\theta}$. The parameters $K_1 = K_2 = 0.02$ are fixed. (a) The monotone type with low sensitivity (switch free). (b) The monotone type with high sensitivity (switch-like). (c) The response curve corresponding to the hysteresis point $\tilde{\Theta}_0$, which is characterized by a vertical tangent. (d) The toggle switch. (e) The response curve corresponding to the irreversibility point $\tilde{\Theta}_1$, which is characterized by $\tilde{I}_{\text{off}} = 0$. (f) The irreversible switch.

to at most one A value (the image). But in the range $[\tilde{I}_{\text{off}}, \tilde{I}_{\text{on}}]$, one $\tilde{I}$ value corresponds to three A values. The middle A value is unstable and is thus represented by a dashed line. The other two A values are stable and are represented by solid lines. The curve is called a toggle switch in the following sense. As the insulin level increases, AKT can be activated once $\tilde{I}$ exceeds the threshold $\tilde{I}_{\text{on}}$. As the insulin level decreases, AKT can be deactivated once $\tilde{I}$ drops below the other threshold $\tilde{I}_{\text{off}}$. Therefore, the state A toggles freely between approximately 0 and approximately 1, and the switching is controlled by realistic insulin concentrations. As $\tilde{\theta}$ increases within the range $\tilde{\Theta}_0 < \tilde{\theta} < \tilde{\Theta}_1$, the response curve changes its shape. In particular, the value of $\tilde{I}_{\text{off}}$ decreases consistently.

Irreversibility point ($\tilde{\theta} = \tilde{\Theta}_1$). The corresponding response curve has $\tilde{I}_{\text{off}} = 0$ (Figure 9.5(e)).

Irreversible switch ($\tilde{\theta} > \tilde{\Theta}_1$). The corresponding response curve has $\tilde{I}_{\text{off}} < 0$ (Figure 9.5(f)), which can no longer be called a toggle switch. It is called an irreversible switch in the following sense. As the insulin level increases, AKT can be activated once $\tilde{I}$ exceeds the threshold $\tilde{I}_{\text{on}}$. However, once activated, it remains activated, even if insulin is completely withdrawn. Indeed, it requires a negative insulin concentration to switch off AKT activation, because the threshold $\tilde{I}_{\text{off}}$ is negative. But a negative insulin concentration is apparently absurd.

Figure 9.5 has provided an informative picture of the deformation of the response curve rendered by the change of $\tilde{\theta}$. As $\tilde{\theta}$ increases, the response curve becomes more and more switch-like, and turns into a toggle switch once $\tilde{\theta} > \tilde{\Theta}_0$. Further increase of $\tilde{\theta}$ turns the toggle switch into the irreversible switch.

9.6.4 The Parameter K and the Limit Condition $K \to 0$

The PI3K-AKT-TOR pathway consists of three components: input, output, and feedback. The output component (colored in blue in Figure 9.1) is the PdPC of AKT, which is a typical example of reversible covalent modification. Albert Goldbeter and Daniel E. Koshland [11] revealed a novel mechanism of sensitivity in reversible covalent modification, known as zero-order ultrasensitivity (see Section 5.5). It can be realized by making the parameter

$$K = \frac{K_m}{X_T}$$

small, where K_m is the Michaelis constant of the enzyme kinetics and X_T is the concentration of the substrate. The smaller K is, the more sensitive the response. To make K small, X_T has to be large. That is, the substrate has to be in a large concentration to saturate the surface of enzymes. Therefore, this condition for generating sensitivity is often called "enzyme saturation."

For the PI3K-AKT-TOR pathway, zero-order ultrasensitivity is conferred if $A_{\max}$, the concentration of total AKT, is large. Indeed, a large $A_{\max}$ makes both

$$K_1 = \frac{K_{m1}}{A_{\max}} \text{ and } K_2 = \frac{K_{m2}}{A_{\max}}$$

small, which confers sensitivity required for the functioning of the PI3K-AKT-TOR pathway.

Because AKT is a well-known oncoprotein (promoting cancer cell growth, proliferation, and survival), it seems natural to reduce $A_{\max}$ (by suppressing the mRNA expression of AKT) for cancer treatment. There are indeed many experimental and clinical practices under way in that direction. This book, however, stresses that a large total AKT concentration is in fact a good thing; it is necessary for sensitive responses that are required by most biological functions. The inhibition of total AKT would inevitably reduce sensitivity and degrade the quality of the pathway response. To inhibit cancer growth, it is necessary to reduce pAKT, but one should not target the total AKT for that purpose. It is more reasonable to prevent AKT activation by targeting other molecules (e.g., PI3K).

The small values of K_1 and K_2 allow one to use the limit condition $K_1 \to 0$ and $K_2 \to 0$ to simplify analysis. By Equations (D.9 and D.12), the limit condition renders $\Theta_0(K_1, K_2) \to 0$ and $\Theta_1(K_1, K_2) \to 1$. One can see this clearly in Figure 9.4, where the two curves $\tilde{\theta} = \tilde{\Theta}_0(K)$ and $\tilde{\theta} = \tilde{\Theta}_1(K)$ intersect the $\tilde{\theta}$-axis (corresponding to $K = 0$) precisely at 0 and 1. The types of response curve can thus be easily defined:

$$\text{Type} = \begin{cases} \text{monotone type} & \tilde{\theta} < 0 \\ \text{toggle switch} & 0 < \tilde{\theta} < 1 \\ \text{irreversible switch} & \tilde{\theta} > 1 \end{cases} \tag{9.39}$$

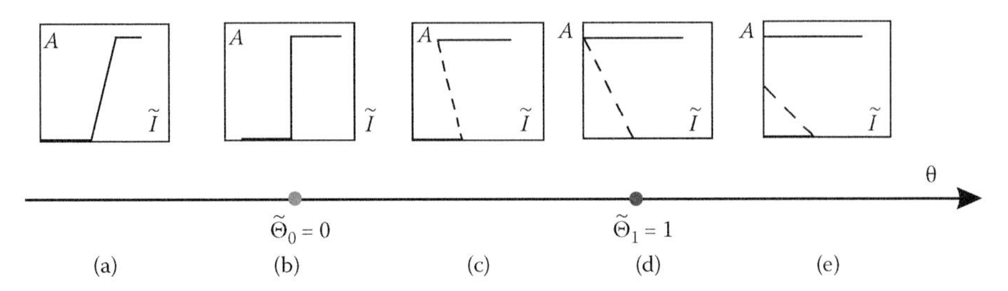

FIGURE 9.6 The deformation of the response curve $A(\tilde{I})$ by the parameter $\tilde{\theta}$, under the condition $K_1 = K_2 = 0$.

In Figure 9.6, response curves of different $\tilde{\theta}$ values are illustrated, under the condition $K_1 = K_2 = 0$. The qualitative transition follows a similar pattern as that in Figure 9.5. It is interesting to note that each response curve is assembled from three straight lines, instead of being a smooth curve.

10 Fundamental Decomposition

Decomposition is a great tool to simplify analysis. For example, the decomposition of force has become a basic technique in physics and is also widely used in other disciplines such as engineering. In linear algebra, LU decomposition factors a matrix as the product of a lower triangular matrix and an upper triangular matrix, which greatly simplifies operations such as the inversion of a matrix, the calculation of the determinant of a matrix, etc. The decomposition of a biological network not only simplifies its analysis, but also allows for a clearer elucidation of the relationship between network structure and biological function [36]. It is thus expected that the decomposition of the PI3K-AKT-TOR pathway would provide valuable insights into the study of metabolism and growth.

The key idea is still to let $K \to 0$. This is a biologically plausible assumption, because a small value of K is indeed required for the pathway to respond sensitively to stimuli. The assumption greatly simplifies the mathematical model. By substituting $K_1 = K_2 = 0$ into form I of the simplified master equation (i.e., Equation (9.35)), one has

$$\alpha\beta A^3 + (-\alpha\beta + \beta\gamma I - 1) A^2 + (1 - \beta\gamma I) A = 0, \tag{10.1}$$

namely,

$$A(A-1)\left(A - \frac{1}{\alpha\beta} + \frac{\gamma I}{\alpha}\right) = 0. \tag{10.2}$$

Equation (10.2) corresponds to three straight lines in the I versus A plane:

$$A = 0, \tag{10.3}$$

$$A = 1, \tag{10.4}$$

$$A = \frac{1}{\alpha}\left(\frac{1}{\beta} - \gamma I\right). \tag{10.5}$$

The three lines intersect at $(I_{\text{on}}, 0)$ and $(I_{\text{off}}, 1)$ (see Figure 10.1). Define $\Delta I = I_{\text{on}} - I_{\text{off}}$ as the *hysteresis width*. One can easily determine their values: $I_{\text{on}} = \gamma^{-1}\beta^{-1}$, $I_{\text{off}} = \gamma^{-1}\beta^{-1} - \gamma^{-1}\alpha$, and $\Delta I = \gamma^{-1}\alpha$. Of the three values,

$$I_{\text{on}} = \gamma^{-1}\beta^{-1} \tag{10.6}$$

and

$$\Delta I = \gamma^{-1}\alpha \tag{10.7}$$

are most illuminating. Assuming γ is a constant, one sees that the switch-on threshold I_{on} depends only on β and has nothing to do with α, while the hysteresis width ΔI depends only on α and has nothing to do with β. That is, by the assumption $K \to 0$, the roles played by α and β are completely decoupled.

It is quite reasonable to assume the constancy of γ, which has the expression

$$\gamma = \frac{k_1}{\delta k_2 E_{2T}}.$$

Indeed, the parameters k_1, k_2, and δ are rate constants, which are relatively constant; E_{2T} represents the total concentration of the phosphotases, which is not controlled directly by insulin and can thus be assumed constant.

Under the condition $\alpha > 0$, the three lines constitute a bistable switch (Figure 10.1(a)). The lines $A = 0$ and $A = 1$ correspond to stable solutions and are thus drawn with solid lines. The line $A = \alpha^{-1}\left(\beta^{-1} - \gamma I\right)$ corresponds to an unstable solution and is thus drawn with a dashed line. This condition renders a positive hysteresis width ($\Delta I > 0$), which is the defining characteristic of hysteresis.

Under the condition $\alpha < 0$, the three lines all correspond to stable solutions; they constitute a monotonically increasing function (Figure 10.1(c)). This condition renders a negative hysteresis width ($\Delta I < 0$), which explains why hysteresis is replaced by monotonicity. In particular, I_{off} loses its alleged meaning (the threshold at which AKT activation switches off); it now corresponds to the insulin level at which the AKT activation level first reaches the maximal.

Under the condition $\alpha = 0$, the three lines constitute a step function because the line $A = \alpha^{-1}\left(\beta^{-1} - \gamma I\right)$ becomes vertical (Figure 10.1(b)). This is not strange because $\alpha = 0$ implies $\tilde{\theta} = \alpha\beta = 0$, which is exactly the hysteresis point under the limit condition $K \to 0$.

10.1 STRUCTURAL DECOMPOSITION

The above analysis reveals the importance of the three parameters K, α, and β. Interestingly, the three parameters correspond to the three components of the PI3K-AKT-TOR pathway.

The parameter β represents the gain of the input component (colored in green in Figure 9.1) of the pathway. Through this component, insulin finally activates the enzyme E_1. Because PI3K is the major molecule of the input component, β can be regarded as the activity of PI3K.

The parameter K characterizes the output component (colored in blue in Figure 9.1) of the pathway, which is the PdPC of AKT $\rightleftharpoons$ pAKT. This can be easily seen from the expression $K = (d + k) / (a A_{\max})$, where d, k, and a are kinetic constants of the PdPC reactions and $A_{\max}$ is the total concentration of AKT, the substrate of PdPC.

The parameter α characterizes the feedback component (colored in red in Figure 9.1) of the pathway, which consists of positive feedback (with the strength Φ) and negative feedback (with the strength Ψ). Note that Ψ reflects TOR activity, which is regulated by many factors such as the intracellular nutrient level, energy status, etc. The parameter α is the difference between positive feedback and negative feedback: $\alpha = \Phi - \epsilon\Psi$.

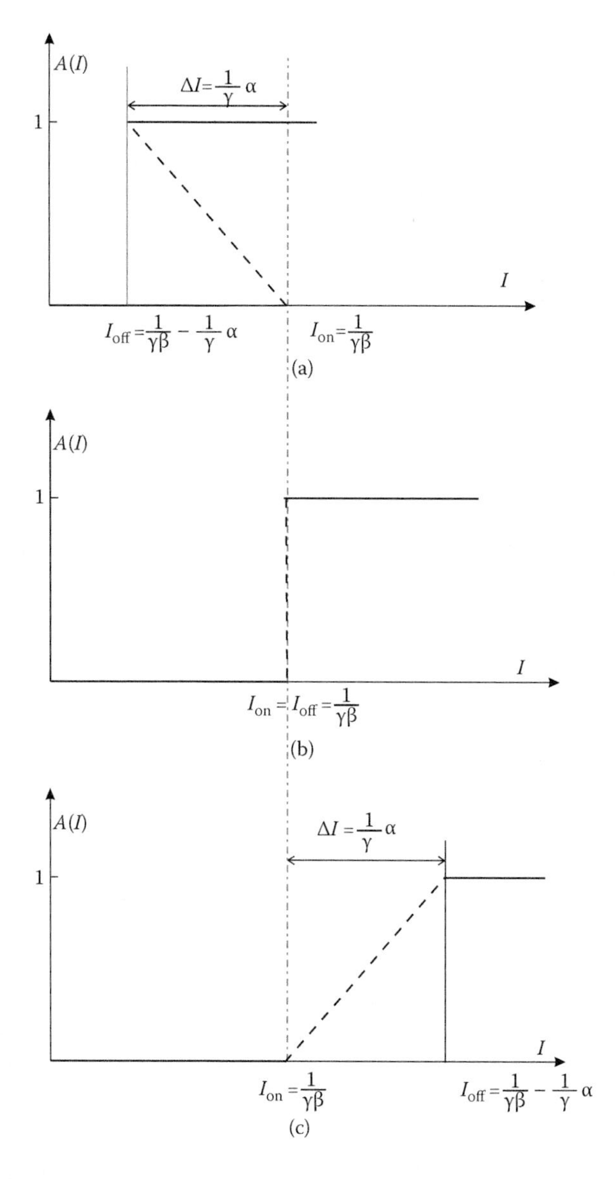

FIGURE 10.1 Under the condition $K = 0$, the three solutions of Equation (9.35) can be represented by three straight lines. (a) With $\alpha > 0$, the three lines constitute a bistable switch ($I_{on} > I_{off}$). (b) With $\alpha = 0$, the three lines constitute a step function ($I_{on} = I_{off}$). (c) With $\alpha < 0$, the three lines constitute a monotonically increasing function ($I_{on} < I_{off}$).

10.2 FUNCTIONAL DECOMPOSITION

The decomposition of the pathway structure implies that the biological functions of the pathway can also be decomposed. Indeed, the three components, through the three parameters K, α, and β, provide distinct biological functions.

10.2.1 The Parameter K

The parameter K confers sensitivity. The smaller K is, the more switch-like the response. Figure 10.2 illustrates many response curves corresponding to different K and α values, but with $\beta = \gamma = 1$ fixed. These curves are grouped according to their K values: $K = 0.00$ (Figure 10.2(a)), $K = 0.01$ (Figure 10.2(b)), and $K = 1.00$ (Figure 10.2(c)).

For the curves with $\alpha < 0$, the increase of K affects the curves only slightly. For example, the gray curve ($\alpha = -3$) does not change dramatically as K increases from 0.00 to 1.00. This is easy to explain — a curve with $\alpha = -3$ is already insensitive even when K is zero, so the increase of K does not further reduce sensitivity too much.

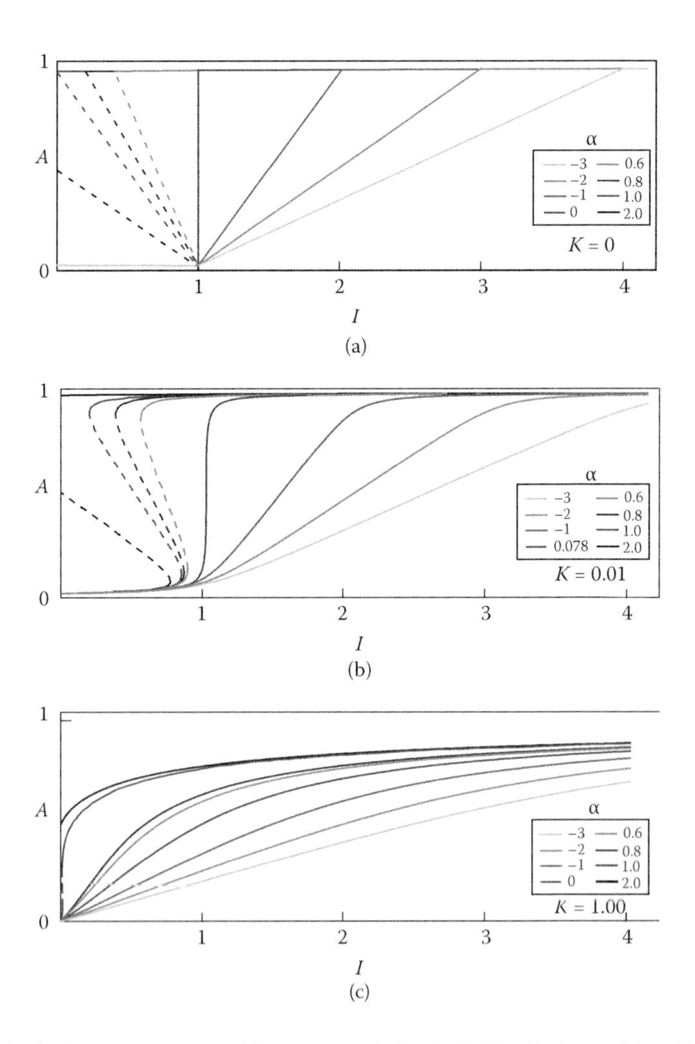

FIGURE 10.2 Response curves with $\beta = \gamma = 1$ fixed. (a) The limit condition $K = 0$. (b) The condition $K = 0.01$. (c) The condition $K = 1.00$. See color insert.

For the curves with $\alpha > 0$, the increase of K reduces sensitivity dramatically. Compare the blue curve in Figure 10.2(b) ($K = 0.01$) with the blue curve in Figure 10.2(c) ($K = 1.00$). The former manifests all-or-none (a desired feature), because the middle values of A (the dashed segment of the blue curve) are unstable and thus do not exist in practice. The latter is dominated by middle values of A. As I increases from 0, A rapidly leaves 0 (undesired). On the other hand, even when I is high, A is still far smaller than 1 (again undesired). All in all, the curves in Figure 10.2(c) are largely hyperbolic, without the desired sigmoidal, switch-like shape.

10.2.2 THE PARAMETER β

The parameter β confers *systemic regulation*. The expression $I_{\text{on}} = \gamma^{-1}\beta^{-1}$ reveals that the threshold of AKT activation I_{on} is determined by β. To better understand this parameter, four β values are used to generate their respective bistable switches, with $K = 0$, $\alpha = \gamma = 1$ fixed (Figure 10.3(a)). As β changes, the bistable switch shifts along the I-axis without any distortions. Indeed, the threshold I_{on} is determined by β (Equation (10.6)) thus the change of β causes the shift of the bistable switch. On the other hand, the hysteresis width ΔI is independent of β (Equation (10.7)), which explains the distortion-free property. Now let $\alpha = -1$, which turns the bistable switches into monotonically increasing functions, but the characteristic (shift without distortion) remains the same (Figure 10.3(b)).

This shift phenomenon can be interpreted either in terms of a single cell (which can respond to changes of metabolic conditions by shifting its response curve along the I-axis) or in terms of four different cells (which each correspond to a response curve in Figure 10.3(a)).

For the first interpretation, the shift mechanism is a good example of zero-order adaptivity defined in Section 5.6. The shift of a cell's response curve can be caused by many factors in the cell, which converge on the input component of the pathway (e.g., by perturbing PI3K activity). This causes the change of β and consequently the shift of both I_{on} and I_{off}, while $\Delta I = I_{\text{on}} - I_{\text{off}}$ remains fixed. This mobile characteristic may confer a certain adaptivity on the cell. That is, the cell can adjust its response to insulin (or other growth factors) according to different metabolic or growth conditions by shifting the response curve along the I-axis. The mechanism can explain many phenomena in biology. For example, muscle cells need to adjust their insulin responsiveness according to different metabolic conditions. Through decreasing the β value, muscles can enhance their insulin responsiveness after exercise or in a cold environment. Through increasing the β value, muscles can settle into a state of insulin resistance in order to reduce glucose uptake. The very existence of I_{on} as a threshold to delay insulin action is already an indicator of the importance of mild insulin resistance as a protective mechanism: muscles do not uptake glucose when the insulin level is smaller than I_{on}, thus sparing glucose for the brain and other more important sites in the body.

Consider a woman in pregnancy. The pre-pregnancy I_{on}, which is sufficient to protect the mother's brain, may now be insufficient to protect the fetal brain and the mother's brain all together (two heads are better than one; but two heads are also more expensive than one). The I_{on} value has to be set higher during pregnancy, and hopefully be restored afterward. This transiently augmented insulin resistance, which underlies gestational diabetes, may well be a natural, useful mechanism to shunt the limited glucose from the mother's muscles and divert it to the fetus. This is particularly important if the mother is afflicted with undernutrition, a common situation among not-too-long-ago human populations. In this event, it might be better that no glucose at all goes to the mother's muscles. Gestational insulin resistance can be induced by the placenta, which secretes placental growth factor to promote the expression of the p85α subunit of PI3K [2]. The p85α subunit competes in a dominant negative fashion with the p85-p110 heterodimer for binding to the IRS protein, which reduces the β value and thus increases I_{on}. In brief, the growth of the fetus is accompanied by an adaptive increase of the I_{on} values of the mother's muscle cells. This adaptation is usually reversible. With the withdrawal of the placental growth factor after childbirth, the original β and I_{on} values should be restored under normal conditions. Zero-order adaptivity is indeed convenient to adjust. And it really protects fetuses!

However, in many populations of modern society, the undernutritional condition is largely removed, which reduces the usefulness of gestational insulin resistance. Compounded with more common conditions such as overnutrition and physical inactivity, the dark side of gestational insulin resistance begins to prevail, contributing to the epidemic of diabetes.

For the second interpretation, the bistable switches in Figure 10.3 (a) may represent the response curves of four different cells. That is, different cells have different values of β, and thus different thresholds of AKT activation and deactivation. The differences might be significant for cells of different types, which creates an order of glucose uptake — more important tissue (with larger β and smaller I_{on}) turns on glucose uptake earlier. This implies a systemic regulation mediated by insulin or growth factors to coordinate metabolism/growth between different tissues. We and other multicellular organisms are composed of many different types of cells. The evolution of multiplicity necessitates novel mechanisms to coordinate growth between different tissues and cell types and to adjust growth according to varying nutrient conditions. Nature solved the problem by two coordinated strategies. First, many growth factors including insulin are developed that can diffuse to virtually everywhere in the body, whereby information about whole-body nutritional and energy status can be conveyed to every cell. Second, different cell types have different I_{on} values, which creates an order of AKT activation, nutrient uptake, and growth. That is, cell types with smaller I_{on} values respond earlier, take up more nutrients, and thus grow faster or store more energy. From this perspective, an adult's myocytes (i.e., muscle cells) may generally have smaller I_{on} values than adipocytes (i.e., fat cells), which have no urgency at all to process glucose. Figure 10.4 shows a possible scenario. Both myocytes and adipocytes are heterogeneous in terms of their I_{on} values. The red and black curves represent the distribution of the I_{on} values of myocytes and adipocytes, respectively. One sees that the mean I_{on} value of myocytes is significantly smaller than that of adipocytes. This renders an ordered glucose uptake: the majority of myocytes uptake glucose first, followed by adipocytes and the other myocytes. If a small amount of meal was ingested, then the peak insulin concentration is so small that no adipocytes are activated to take in glucose (Figure 10.4(a)). If a large amount of meal was ingested, then some adipocytes are activated to take in glucose and convert it into fat for storage (Figure 10.4(b)).

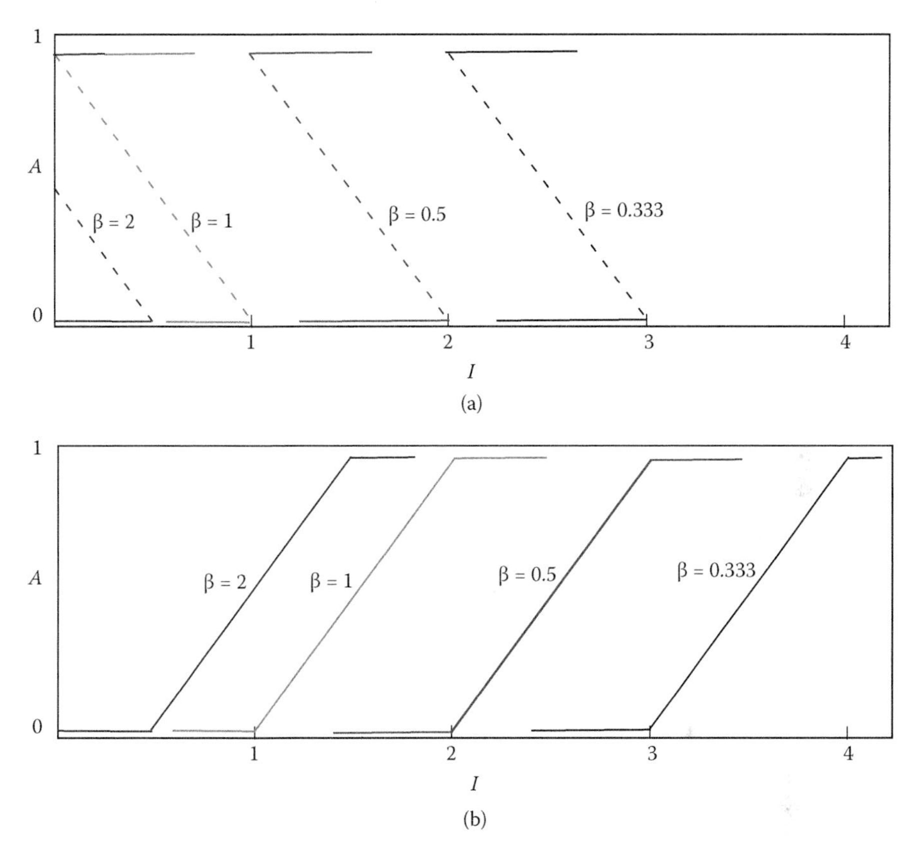

FIGURE 10.3 Response curves with different β values but with $K = 0$ and $\gamma = 1$ fixed. The curves shift along the I-axis without distortion, due to the change of β. (a) The case $\alpha = 1$. (b) The case $\alpha = -1$.

10.2.3 The Parameter α

The parameter α reflects the autonomous aspect of a cell's energy uptake. In the expression $\alpha = \Phi - \epsilon\Psi$, the parameter Φ is the gain from pAKT to IRS (which measures the ability of pAKT to stimulate IRS). It is relatively constant. The parameter Ψ represents the activity of TOR, a protein complex sensing the *intracellular* nutrient concentration. Therefore, α is determined solely by the nutritional state of the cell itself. A more "hungry" cell has a smaller Ψ and thus a larger α.

It is expected that a more hungry cell should manage to absorb more glucose for compensation. This is indeed the case. According to the expression $\Delta I = \gamma^{-1}\alpha$, a cell with larger α has a larger ΔI and thus a longer duration of glucose uptake. Therefore, a more hungry cell absorbs glucose for a longer time. Consider three cells with the same β value (thus the same I_{on}) but different α values (thus different ΔI). Their response curves are drawn in green, blue, and red, respectively, in Figure 10.2(b). In particular, the red response curve belongs to the most hungry cell. It has the largest ΔI, poised to accept more glucose than the other two cells. In other words,

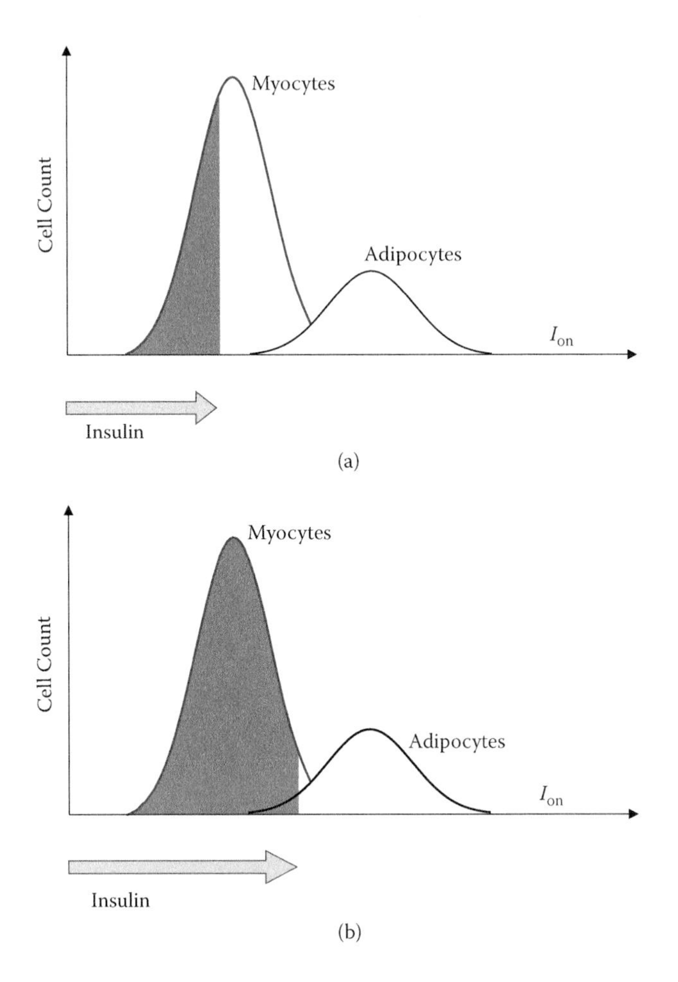

FIGURE 10.4 Myocytes are heterogeneous in terms of their I_{on} values; their distribution is represented by the red curve. Adipocytes are also heterogeneous in terms of their I_{on} values; their distribution is represented by the black curve. The yellow arrow represents the peak insulin level. (a) The peak insulin level is smaller than all the I_{on} values of adipocytes; thus no adipocytes are activated. The portion of activated cells is illustrated in green. (b) The peak insulin level is larger than the I_{on} values of a small portion of adipocytes. These adipocytes are thus activated (indicated by the small region colored in green underneath the black curve). See color insert.

the three cells begin to absorb glucose simultaneously (having the same I_{on}), but the most hungry cell (corresponding to the red curve) is the last one to switch off glucose uptake.

This mechanism allows a cell to adjust its glucose uptake according to the level of nutrients that has already been acquired. The plausibility of this view is supported by the fact that TOR is an ancient protein complex that is even possessed by unicellular organisms such as yeast. Unlike multicellular organisms, which require complex regulations to enforce a coordinated growth, unicellular organisms do not. Each cell is an individual organism. It takes up nutrients according to its own needs, with no requirements for coordination with others. The parameter α thus corresponds to the cell's instinct for ingestion.

10.2.4 SUMMARY

The structure-function relationship of the PI3K-AKT-TOR pathway has now become clear. The pathway can be divided into three components (input, feedback, and output). The input component, characterized by the parameter $\beta > 0$, imposes a global control upon growth and metabolism, mediated by systemic growth factors such as insulin. The larger β is, the earlier the switching on of glucose uptake. That is, cells with smaller β have to wait for a while to begin glucose uptake, during which cells with larger β have already been absorbing glucose. The feedback component, characterized by the parameter $\alpha > 0$, realizes a local (cell autonomous) control of nutrient uptake. The smaller α is, the earlier the turning off of glucose uptake. That is, cells with larger α are more avid, poised to get more glucose. The output component, characterized by the parameter $K > 0$, confers sensitivity. The smaller K is, the more sensitive the response to growth factors.

One can consider the fundamental decomposition as a major advancement of our understanding of the regulatory mechanisms of the PI3K-AKT-TOR pathway. The possession of an entire PI3K-AKT-TOR pathway distinguishes multicellular organisms from unicellular organisms. TOR, an ancient molecule that is even possessed by unicellular organisms, mediates nutrient uptake in a cell in an autonomous manner. The evolution of multicellular organisms necessitates novel mechanisms to coordinate growth between tissues and to adjust growth according to varying nutrient conditions. Nature solved this problem by wiring TOR into a circuit that is now known as the PI3K-AKT-TOR pathway, which brought cell autonomous ingestion under whole-organism systemic regulation. A less important cell, even if hungry, has to wait until more important cells (with smaller I_{on} values) have already begun glucose uptake. Neurons are the most important, and thus their glucose uptake is independent of insulin ($I_{\text{on}} = 0$). Therefore, the brain always utilizes glucose actively (as it needs) even if the plasma glucose concentration is very low. In contrast, the muscles have to wait until the plasma glucose concentration is high enough to guarantee the brain's safety. Adipose tissues have to wait even longer, because their primary role is to store surplus glucose and it is perfectly fine that they do not get glucose at all.

Moderate changes of α and β confer adaptivity, which is important for an organism to cope with varying metabolic conditions. If their changes do not follow in a timely manner with the varying metabolic conditions, diseases may occur. For example, α may need to become larger under certain extreme conditions. It is expected that the original α value is restored when the metabolic condition becomes normal again. If α sticks at the larger value for some reason, then diseases might develop. Indeed, as will be shown in the following chapters, overly large α or β values would increase the chance of cancer development; overly small α or β would increase the chance of developing type 2 diabetes; an overly large K may also lead to some pathologic consequences.

11 Normal Phenotype

In Chapter 9, mathematical analysis of the PI3K-AKT-TOR pathway revealed three kinds of response curves: irreversible switch, toggle switch, and monotone type. The three varieties occupy three regions in the phase diagram K versus $\tilde{\theta}$ (Figure 9.4). A natural question is: which kind of response corresponds to the typical normal physiology? This question can be addressed by integrating data obtained at multiple biological levels: the evolutionary scale, the organismal level, and of course, the molecular level. These considerations together suggest that under normal conditions, the PI3K-AKT-TOR pathway operates as a toggle switch. In brief, normality corresponds to bistability.

11.1 EVOLUTIONARY SCALE

Cancer, obesity, and diabetes are the defining plagues of our generation. It has become clear that their epidemics are escalating in recent years. And the *simultaneous* escalation suggests an evolutionary underpinning for cancer, obesity, and diabetes. Being closely related to these diseases, it is quite possible that the biological functions of the PI3K-AKT-TOR pathway have been shaped by evolution.

An eternal theme in evolution is the struggle for life. Natural selection forces an organism to be well suited to the local environment it evolves in (survival of the fittest). Over the course of evolution, natural selection has led to an astounding variety of organisms on our planet. When resources become scarce, different species compete for food in an ecosystem. For example, cheetahs and lions often compete, since both species feed on the same prey, and each can be negatively impacted by the presence of the other because they will have less food. The direct antagonism between prey and predator can also be regarded as a form of competition, with the loser becoming the food of the winner. Those species successful in competition have more chance to survive and pass on their genes (survival of the strongest).

Incessant competition is a driving force for organisms to become more and more complex. The emergence of multicellularity, which marked the differentiation of biological functions, is a key event in the evolution of life. As species become more and more complex, different types of cells constitute different tissues and organs, which are organized to perform a variety of physiological functions. The intense food competition often led to an evolutionary arms race between prey and predator. Although the predator became more competent, the prey evolved new tricks to evade attack. For almost every species, life never becomes easy.

Skeletal muscle is obviously an important weapon for competition. A strong muscle system confers crucial survival advantages on animals. Both predators and prey need strength and fast speed, which can be generated through contraction of muscles. However, muscles and other tissues must be innervated and coordinated to carry out their functions. The specific organ that exerts centralized control is the brain. The brain acts on the rest of the body both by generating patterns of muscle activity and by driving secretion of hormones. It allows groups of muscles to be co-activated in complex patterns; it also allows stimuli impinging on one part of the body to evoke responses in other parts, and it can prevent different parts of the body from acting at cross-purposes to each other. This centralized control allows rapid and coordinated responses to changes in the environment, such as the approach of a natural enemy.

The brain is composed primarily of two broad classes of cells: neurons and glial cells. Each neuron is connected by synapses to several thousand other neurons. These neurons communicate with one another by means of long protoplasmic fibers called axons, which carry trains of signal pulses called action potentials to distant parts of the brain or body, targeting specific recipient cells. These signal pulses last less than a thousandth of a second and travel along the axon at speeds of 1–100 meters per second. Some neurons emit action potentials constantly, at rates of 10–100 per second, usually in irregular patterns; other neurons are quiet most of the time, but occasionally emit a burst of action potentials.

Continued evolution makes the brain more and more complex. For humans, the cerebral cortex constitutes most of the brain. This thick layer of neural tissues is folded in a way that increases the amount of surface that can fit into the volume available. The cerebral cortex is estimated to contain 15–33 billion neurons, each connected by some 100 trillion synapses to several thousand other neurons.

Evolution also stimulates the brain to become bigger. As animals get bigger, so do their brains. Large animals such as whales and elephants may have larger brains in absolute terms, but when measured using the encephalization quotient which compensates for body size, the human brain appears to be the largest — about seven times bigger than that of other similarly sized animals. In the past two million years, the hominid lineage leading to modern humans evolved significantly larger and more sophisticated brains than other primates. The gorilla, which can grow to be three times bigger than us, has a smaller brain than we do. The human brain is about 3.5 times larger than that of chimpanzees (*Pan troglodytes*), which are similar in lean body weight to us. The modern human brain is also approximately three times larger than in the now extinct but smaller very early human ancestors, the Australopithecines. The modern human brain is now approximately 6% smaller than it was in the early modern homo sapiens of 25,000–100,000 years ago, but we are still equally "encephalized" because our body weight is also a little less than theirs.

11.2 ORGANISMAL LEVEL

Competition is an eternal theme in evolution. An animal has to compete with other animals for resources. Even if there were no external competition, the animal would often face food scarcity in an ecosystem. An animal can usually obtain little food, barely enough for its survival. In this event, a careful distribution of nutrients to different parts of the body is crucial for the survival of the animal. If not regulated systemically, cells would naturally compete with each other for the limited nutrients. A regulatory mechanism must be in place to ensure thrift in energy utilization — more important organs should have priority in obtaining nutrients over less important organs. As organisms become more complex, the internal competition becomes more intense and complex, and the internal regulation becomes more difficult. This internal competition thus exerts selection pressure on the animal under discussion. The evolution of an organism is thus shaped by both external and internal competitions.

The human is the most advanced organism in the world. The human body consists of many systems, including the musculoskeletal system, cardiovascular system, digestive system, endocrine system, integumentary system, urinary system, lymphatic system, immune system, respiratory system, nervous system, and reproductive system. Each system consists of several organs, a collection of tissues joined in a structural unit to serve a common function. The organs of the body are composed of four different primary tissues: muscle, nervous, epithelial, and connective tissues, each of which has its own characteristic structure and function. The activities and interactions of these tissues determine the physiology of the organs. Each tissue consists of cells, which can be grouped into different types. For example, nervous tissue consists of neurons and glial cells.

The cells in the body all require energy to sustain their routine activities: biosynthesis, ionic and neutral pumping, mechanical work, etc. Therefore, each cell type should have some basal level of energy utilization. This basal energy supply can be delivered to the cells through GLUT1, a ubiquitous glucose transporter that is expressed in the membrane of virtually every cell. GLUT1 has a small Michaelis constant of $K_m \approx 1.5$ mM, far smaller than the usual plasma glucose concentration. This small K_m allows for an essentially constant flow of glucose to every cell, even when the plasma glucose concentration is as low as that of the fasting state (in which the plasma glucose concentration of 4–6 mM is still threefold higher than the K_m of GLUT1). Therefore, GLUT1 is almost always saturated with glucose and operates near a constant rate V_{max}. However, this democratic distribution of glucose alone does not meet the complex physiological functions of the body. Just as in a society, whereas a basal supply should be delivered to all people to ensure their survival, certain polarity and differentiation are allowed for the vigorous functioning of the society. In the body, cells have a great heterogeneity in glucose utilization. Glucose is indispensable for some organs and some organs have alternative fuels. Some cells are energy expensive and some cells utilize little energy. The brain and muscles follow quite different mechanisms in glucose utilization.

11.2.1 Brain

The brain has the supreme priority in energy utilization, among all the organs in the human body. Being extraordinarily large, the human brain consumes a great deal of glucose. However, the size alone does not explain the energy actually utilized by the brain. The brain weighs approximately 2.5% of the body weight, but it uses approximately 25% (a ten times difference!) of the body's daily energy requirement. This great discrepancy is due to complex interconnections between neurons, upon which a great deal of information is processed through the transmission of action potentials. There are 15–33 billion neurons, each connected by some 100 trillion synapses to several thousand other neurons. Some neurons emit action potentials constantly, at rates of 10–100 per second, which last less than a thousandth of a second and travel along the axon at speeds of 1–100 meters per second. These neural activities (the maintenance of the Na^+–K^+ membrane potential and the transmission of nerve impulses) demand a great deal of energy cost.

Because fatty acids cannot cross the blood–brain barrier, the brain depends absolutely on glucose as its sole energy source. Moreover, neurons do not store glucose (unlike muscles, which can store glucose in the form of glycogen). Hence the brain requires a *continuous* supply of glucose to support the bustle of neurons. Every day a human brain consumes about 120 g of glucose, which corresponds to an energy value of about 1760 kJ (420 kcal). This accounts for some 60% of glucose utilization by the whole body in the resting state. Even when the total energy (including fatty acids) is considered, the brain still consumes about 25% of total energy, which is more than any other organ. To satisfy this unique glucose requirement, the highly efficient glucose transporter GLUT3 is expressed in the brain. GLUT3 is the major neuronal glucose transporter, presented in both dendrites and axons of neurons. GLUT3 has two extraordinary properties. First, GLUT3 has an extremely high affinity with glucose. Its K_m value is the lowest among all the GLUTs. Such a high affinity is crucial for the brain. Due to the blood–brain barrier, the ambient glucose concentration surrounding the neurons is only 1–2 mM compared with 4–6 mM in serum. Second, GLUT3 also has an extremely high transport capacity. The turnover number k_{cat} of GLUT3 is the highest among all the GLUTs (at least fivefold greater than GLUT1). The high ratio k_{cat}/K_m implies that GLUT3 is a highly efficient transporter, which allows for a relatively constant rate of glucose uptake even at small plasma glucose concentrations (Figure 11.1(a)).

11.2.2 LIVER

The absolute glucose dependence of the brain necessitates a constant source of endogenous glucose. Because food ingestion is sporadic, the body must have an organ to store glucose when the plasma glucose concentration is high and to release glucose when the plasma glucose concentration is low. The specific organ that performs the task is the liver. Muscles can store but cannot release glucose. Once glucose enters a myocyte, it is phosphorylated and is thus "trapped" in the cell. On the other hand, adipocytes turn glucose into fat and release it as fatty acids, but not glucose.

11.2.3 MUSCLE

Muscles constitute about 50% of the whole body mass and utilize a great deal of glucose. However, glucose metabolism in muscles is not as crucial as it is in the brain. First, muscles are far less important than the brain. The removal of some bulk muscles won't jeopardize one's life, but that simply does not apply to the brain. Second, glucose utilization by muscles is optional. Muscles preferentially burn fatty acids under most physiological conditions. Therefore, a tonic glucose uptake by muscles is unnecessary, and the scheme illustrated in Figure 11.1(a) is not suitable for muscles. If the scheme were used by muscles, then the brain would be deprived of glucose when the plasma glucose concentration is low, because the 2.5% body mass simply cannot compete with the 50% body mass. It is more reasonable to follow the scheme illustrated in Figure 11.1(b). When the plasma glucose concentration is low (high), the rate of glucose uptake is small (large). This can protect the brain to some degree, but is insufficient, as revealed by Figure 11.1(b). When the plasma glucose concentration is at a small value G_1, a myocyte still uptakes glucose at an observable rate kG_1, where

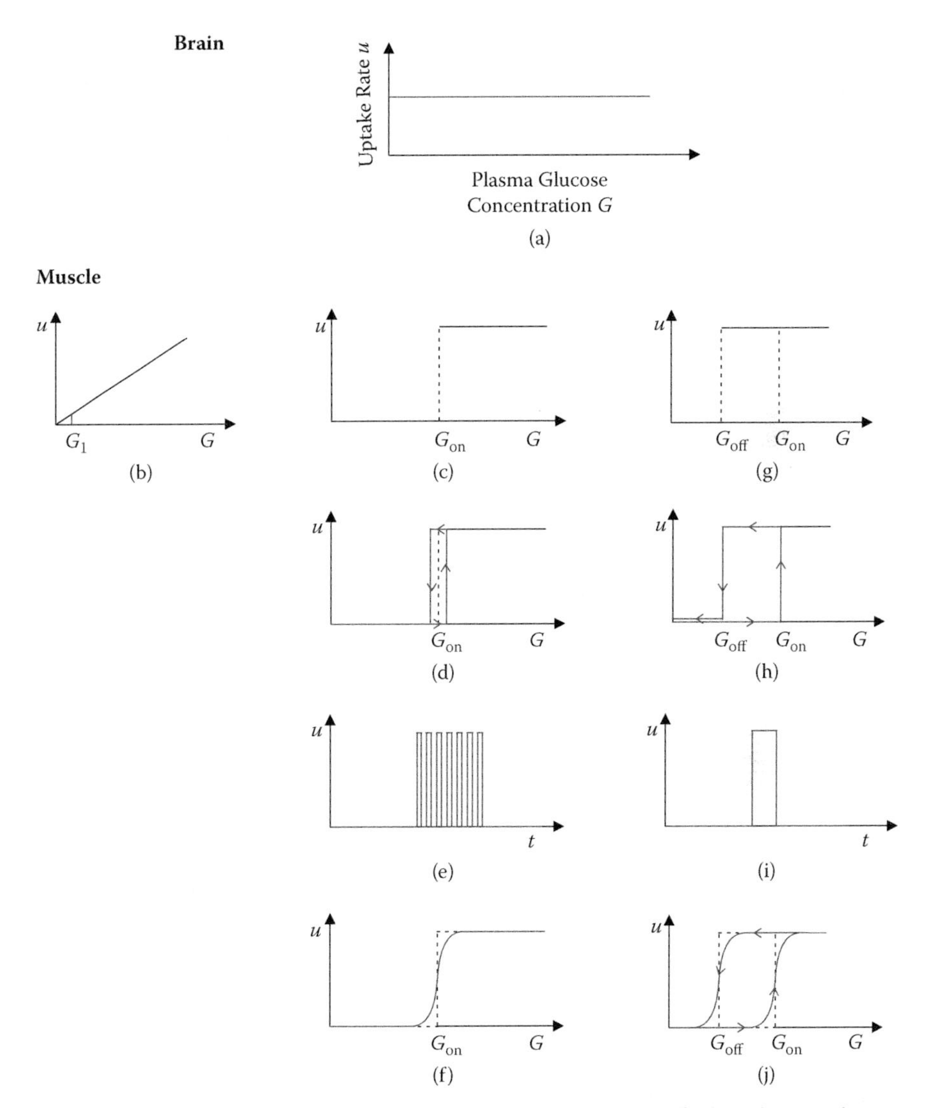

FIGURE 11.1 The rate of glucose uptake u as a function of the plasma glucose concentration G. (a) The brain has a relatively constant rate of glucose uptake. (b) Putative pattern of muscular glucose uptake: a gradual increase as G increases. (c) Putative pattern of muscular glucose uptake: a step function that switches at G_{on}. (d) A reproduction of (c), with the clattering phenomenon highlighted in red. (e) Putative muscular glucose uptake as a function of time, according to (c). (f) A modification of (c), with the effect of cell heterogeneity considered. (g) Putative pattern of muscular glucose uptake: a toggle switch that turns on at G_{on} and turns off at G_{off}. (h) A reproduction of (g), which highlights in red the sensitive and robust response. (i) Putative muscular glucose uptake as a function of time, according to (g). (j) A modification of (g), with the effect of cell heterogeneity considered. See color insert.

k is the slope of the response. Although the diversion is small to a single myocyte, the cumulative diversion is large because muscles are so massive. The disastrous consequence would be that muscles drain the precious plasma glucose that should be spared for the brain.

To solve this problem, muscles can adopt an all-or-none strategy (Figure 11.1(c)). When the plasma glucose concentration is low ($G < G_{\text{on}}$; e.g., in the fasting state), muscles do not take up glucose so that the precious glucose can be spared for the brain. During this period, muscles use fatty acids and energy stores (glycogen and intramuscular triglyceride) to provide energy. When the plasma glucose concentration is very high ($G > G_{\text{on}}$; e.g., immediately after a meal), muscles absorb glucose with the full force to clear plasma glucose as soon as possible (i.e., to avoid hyperglycemia).

However, this simple scheme may not even work. Consider what would happen after a meal. The plasma glucose concentration increases rapidly and soon exceeds the threshold G_{on}, which triggers muscles to absorb glucose. Because each myocyte works with full force and there are so many myocytes, G stops increasing almost immediately and then decreases rapidly. That is, once G reaches G_{on}, it immediately drops below G_{on}. According to the scheme (Figure 11.1(c)), muscles switch off glucose uptake immediately. Our story does not end here though. Because food remaining in the digestive tract continues to supply glucose, the plasma glucose concentration again increases and exceeds G_{on}, which again triggers muscles to absorb glucose, according to the scheme. But this switch-on state of muscles does not last — they have to switch off almost immediately, as already explained. The cycle repeats infinitely many times, if the scheme is exactly all-or-none. In reality, G oscillates with high frequency around G_{on} many times (the number of times depends on the amount of the meal), until almost all the food has been consumed. This oscillation behavior is illustrated by the red "clattering" cycle in Figure 11.1(d). It is also illustrated as a time function $u(t)$, namely, the square wave with many pulses in Figure 11.1(e).

It is here that hysteresis can rescue us. To avoid oscillation (or at least reduce the frequency of oscillation), myocytes can postpone switching off glucose uptake by using a threshold G_{off} that is smaller than G_{on} (Figure 11.1(g)). Consider again what would happen after a meal. The plasma glucose concentration increases rapidly and soon exceeds the threshold G_{on}, which triggers muscles to absorb glucose. Because muscles are so massive, G decreases immediately and is soon below G_{on}, the same as above. Now there is a difference. Here muscles continue to uptake glucose with full force due to hysteresis, until G drops below the threshold G_{off}. This hysteresis width $\Delta G = G_{\text{on}} - G_{\text{off}}$ thus confers robustness of muscular glucose uptake. If ΔG is large enough, the oscillation can be avoided completely — one cycle is sufficient for glucose homeostasis (Figure 11.1(h)). In this event, the corresponding time function $u(t)$ is a single pulse with a stable duration (Figure 11.1(i)).

The above reasoning was based on the assumption that all the myocytes respond simultaneously (i.e., the cells have the same G_{on} and G_{off} values). In practice, there exists a certain heterogeneity among myocytes in terms of their G_{on} and G_{off} values. Therefore, total muscular glucose uptake might follow a kinetic curve such as the one shown in Figure 11.1(j): not exactly all-or-none but the hysteresis characteristic is still salient.

11.2.4 Pancreas and Insulin

The scheme illustrated in Figure 11.1(g) depends on the plasma glucose concentration as the control signal to switch muscle actions on or off. Moreover, the scheme requires glucose to realize a sensitive switching. A natural question is: how does glucose realize signaling? In other words, how does a cell detect the rise and fall of glucose in the blood so that it can respond accordingly? Signal transduction is usually conducted through the interaction between two molecules, e.g., a peptide hormone and its receptor on the cell surface. It is required that both molecules have sufficiently complex, well-matched conformations for the purpose of recognition. Glucose, a small molecule with a molar mass only 180 g/mol, has a structure too simple to act as an efficient signaling molecule. Therefore, G is actually not suitable to act as the horizontal axis in Figure 11.1(g). Among biomolecules, proteins (including peptides) have complex three-dimensional conformations and are commonly used as signaling molecules.

Insulin is a peptide hormone with a molar mass of 5808 g/mol, which is complex enough for molecular recognition and signaling. Accordingly, there are insulin receptors on the cell surface to interact with insulin and to detect the change of its concentration. To substitute glucose as the signaling molecule, the plasma insulin concentration must track closely with the plasma glucose concentration. Only with that can one safely replace G with I, G_{on} with I_{on}, and G_{off} with I_{off} in Figure 11.1(g).

In the body, insulin is produced by the pancreas in response to the stimulus from plasma glucose. The higher the glucose concentration, the more insulin is secreted, and the higher the insulin concentration. Therefore, the rise and fall of insulin should in principle follow closely with the rise and fall of glucose. Indeed, clinical data proved that this is the case. Figure 3.2(f) shows the 24-h profiles of plasma glucose concentration, averaged from 14 normal and 15 obese persons. The three spikes are caused by the three meals. Figure 3.2(e) shows the corresponding insulin profiles. One sees that the insulin changes follow the same pattern as the glucose changes.

In summary, muscular glucose uptake $u(I)$ should be a bistable response to the plasma insulin concentration I (Figure 11.1(g) with the horizontal axis replaced by insulin). The response is characterized by all-or-none and hysteresis. In the fasting state, the plasma glucose concentration is small, and the insulin concentration is also small. This prevents muscles from uptaking glucose, and the very limited glucose is thus spared for the brain. After a meal, the plasma glucose concentration increases abruptly, and the plasma insulin concentration increases accordingly. Once I exceeds the threshold I_{on}, muscle cells can then greedily uptake glucose without worrying about brain safety. As a consequence, the plasma glucose concentration decreases, which is paralleled by the fall of the insulin concentration. Muscles stop uptaking glucose only after I drops below I_{off}. If the hysteresis width $\Delta I = I_{\text{on}} - I_{\text{off}}$ is not large enough, then glucose and insulin concentrations may rebound, because food remaining in the digestive tract continues to supply glucose. If the hysteresis width is large, then the remaining food is insufficient to render rebound, and glucose and insulin concentrations continue to reduce to their basal levels.

11.2.5 Optimal Control Perspective

Although discussions at the organismal level have only used intuitive arguments, the conclusion is in line with that obtained by rigorous mathematical analysis in Chapter 3. By using an optimal control perspective, it was demonstrated in Chapter 3 that bistability (toggle switching) is a fundamental mode of homeostatic regulation. When applied to glucose homeostasis, this implies that the optimal distribution of glucose to different organs depends on a bistable response of muscular glucose uptake to insulin stimulus. Mathematical analysis also showed that the optimal control, if represented as a function of time, is none other than a square pulse (Figure 3.2(a)). We therefore have both intuitive arguments and rigorous analysis at the organismal level, and the two match each other seamlessly.

11.3 MOLECULAR LEVEL

Considerations at the organismal level (by both mathematical analysis and intuitive reasoning) conclude that bistability renders optimal glucose homeostasis and it should be the normal mode of regulation. In Chapter 9, it was demonstrated that the PI3K-AKT-TOR pathway, as a master regulatory network controlling many things including glucose homeostasis, can indeed produce a toggle switch as one kind of response to insulin stimulus. Therefore, what is desired at the organismal level can be provided at the molecular level. In the following, we show that considerations at the molecular level alone also lead to the same conclusion, without borrowing information from the organismal level.

Here we use form II of the simplified master equation to address the question, under the limit condition $K \rightarrow 0$ so that analysis can be further simplified. The condition renders a simple classification of response curves: $\tilde{\theta} < 0$ corresponds to the monotone type; $0 < \tilde{\theta} < 1$ corresponds to the toggle switch; $\tilde{\theta} > 1$ corresponds to the irreversible switch.

To establish that the toggle switch corresponds to normality, one has to rule out the cases $\tilde{\theta} > 1$ and $\tilde{\theta} < 0$ as normal responses. In the next chapter we will learn that the irreversible switch ($\tilde{\theta} > 1$) generated by the PI3K-AKT-TOR pathway is a typical pathologic phenotype. Therefore, we just need to show that the normal response has $\tilde{\theta} > 0$. This condition can be proved by considering the expression

$$\tilde{\theta} = \alpha\beta = (\Phi - \epsilon\Psi)\gamma A_{\max}\beta \tag{11.1}$$

together with some well-established biological data. The parameter β is the gain of the input component, which can be represented by the PI3K activity. It is well known that PI3K overactivity promotes cancer. PI3K overactivity can be induced by two kinds of gene mutations. On one hand, the mutation of the gene *PIK3CA* is a gain-of-function one, which enhances PI3K activity. On the other hand, PI3K is negatively regulated by the gene *PTEN*, whose deletion causes overactivity of PI3K. That is, both gene mutations cause PI3K overactivity and promote carcinogenesis. These data reveal the positive correlation between β (whose increase corresponds to enhanced

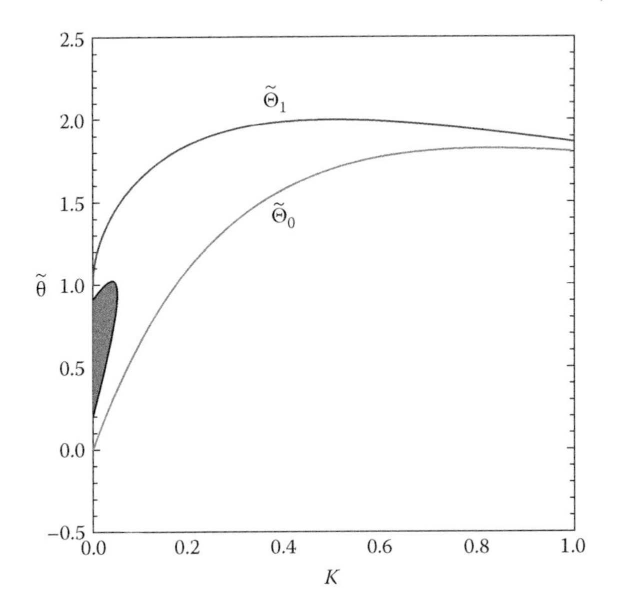

FIGURE 11.2 The phase diagram K versus $\tilde{\theta}$, where the rough region of normality is colored in green. See color insert.

PI3K activity) and $\tilde{\theta}$ (whose increase corresponds to increased cancer risk; see the next chapter). Because $\tilde{\theta} = \alpha\beta$, one must have

$$\alpha > 0$$

to render the positive correlation between $\tilde{\theta}$ and β. According to Equation (11.1), this implies

$$\Phi > \epsilon\Psi.$$

That is, the positive feedback must be stronger than the negative feedback under normal conditions. If Φ were smaller than $\epsilon\Psi$, then an increase of β would make $\tilde{\theta}$ even more negative, which leads to the unreasonable conclusion that the increase of PI3K activity would reduce cancer risk. That is, the presumption $\Phi < \epsilon\Psi$ is contradictory to the established data that β positively correlates with cancer risk. Therefore, under normal conditions, the positive feedback must be stronger than the negative feedback ($\Phi > \epsilon\Psi$), namely, $\tilde{\theta} > 0$.

In summary, normality corresponds to the condition $0 < \tilde{\theta} < 1$ when $K \to 0$. Because in reality K is not zero, the normal phenotype should occupy a narrow region (colored in green in Figure 11.2) that is close to the vertical axis and that is between the red and green curves. The normal response is a toggle switch.

12 Disease Phenotypes

Clinical and epidemiological studies have revealed that many complex diseases are intimately connected. A typical example is the connection between cancer and diabetes. Patients with diabetes and those who are obese have a greater incidence and aggressiveness of cancer development. Conversely, patients with diabetes who are treated with metformin to lower plasma insulin levels have reduced chances of cancer in comparison to untreated individuals. The deep connection between cancer and diabetes is also implied by their simultaneous epidemics in recent years.

The molecular basis of the connection was recently established by gene expression profiling that identified a common set of differentially expressed genes. However, it was not well understood how the integration of individual molecular differences can lead to the emergence of different diseases. From the perspective of systems biology, it is the complex network of interactions among the genes and gene products that leads to the physiologic/pathologic phenotypes that we observe. To deepen our understanding of complex diseases and their mechanistic connections, it is important to uncover system-level control mechanisms of the underlying network.

It is known that malfunctions of the PI3K-AKT-TOR pathway are implicated in diseases related to metabolism and growth. Because the pathway has only a few kinds of response (the monotone type, the toggle switch, and the irreversible switch), as revealed in Chapter 9, it is reasonable to interrogate each kind of response curve to see whether or not it corresponds to a disease phenotype. In the last chapter, we found the correspondence between the toggle switch and normality. In this chapter, we aim to determine if these response curves (including the toggle switch) can potentially lead to diseases.

It should be stressed that conclusions in this chapter are largely relative. When we make the statement "a response curve corresponds to a disease," we mean that the response curve has certain characteristics that increase the risk of the disease, under normal physiologic conditions. Under some extreme conditions, however, the same response curve may become a necessary response and is thus healthy. It is also possible that the response curve is fine for one cell type but is pathologic for another, or is fine only at certain stages of development but not at others.

12.1 CANCER

In Figure 12.1, the red curve ($\alpha = 2$) is an irreversible switch, because the threshold I_{off} is negative and is thus unreachable. The irreversible switch may correspond to the cancer phenotype. Because cancer involves many different growth factors, in the following discussions the symbol I represents a general hormone that can activate the PI3K-AKT-TOR pathway, which includes insulin, insulin-like growth factor, epidermal growth factor, and so on.

Suppose initially the concentration of the growth factor is zero ($I = 0$) and AKT is not activated ($A = 0$). As I increases over the threshold I_{on}, the protein kinase AKT is activated ($A = 1$). Remarkably, this activation becomes permanent in the sense

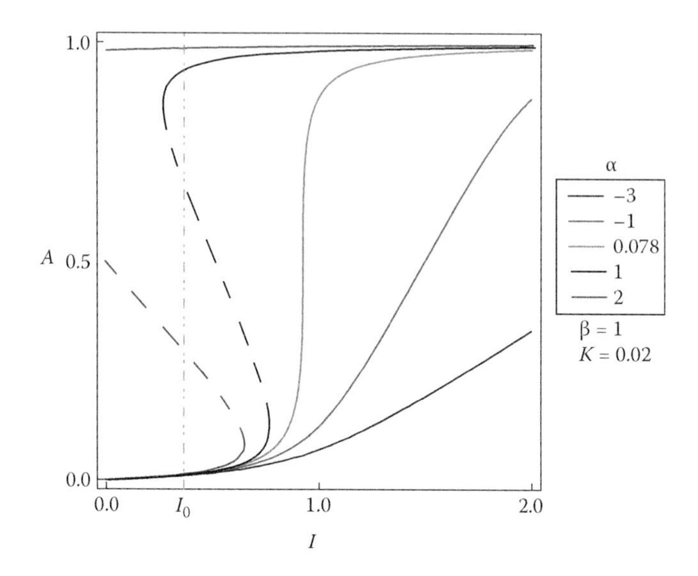

FIGURE 12.1 Response curves $A(I)$ that may correspond to physiologic/pathologic phenotypes. These curves are parameterized by the same values of β and K, but different α values. I_0, indicated by the dash-dot-dot line, represents the basal level of the growth factor I. See color insert.

that it is no longer controlled by the growth factor. To deactivate AKT, the growth factor concentration has to be negative (to make I even smaller than I_{off}), which is impossible. In this event, a transient stimulus gives rise to constitutively active AKT. That is, a single pulse of the growth factor can trigger the permanent activation of AKT, as long as the peak of the pulse is over I_{on}. This activation pertains even if a pulse never arrives again.

It is well known that AKT is an oncoprotein, whose activation promotes carcinogenesis from many aspects. It promotes massive and continuous glucose uptake, which is necessary for cancer metabolism and uncontrolled cell proliferation. Indeed, aberrant metabolism and uncontrolled proliferation are both hallmarks of cancer. In order to become rapidly dividing, cancer cells need to utilize a great deal of glucose. What makes matters even worse is that cancer cells also *waste* a great deal of glucose (the Warburg effect). Therefore, similar to neurons, cancer cells virtually require a continuous supply of glucose. But unlike neurons, cancer cells generally do not have the highly efficient GLUT3. They depend on the constitutive activation of AKT (and/or other oncoproteins) to secure a continuous glucose supply. AKT activation promotes cell survival through coupling with other pathways such as NFκB. Even if the local environment has a very low glucose concentration, the cell may still manage to survive. AKT activation also promotes metastasis through decreasing transcription of the transmembrane protein E-cadherin that forms adhesions between adjacent cells, thereby permitting their detachment and migration to a new environment that is richer in nutrients.

A toggle switch may also lead to the cancer phenotype if its threshold I_{off}, although positive, is overly small. This is because the growth factor usually has a basal level I_0

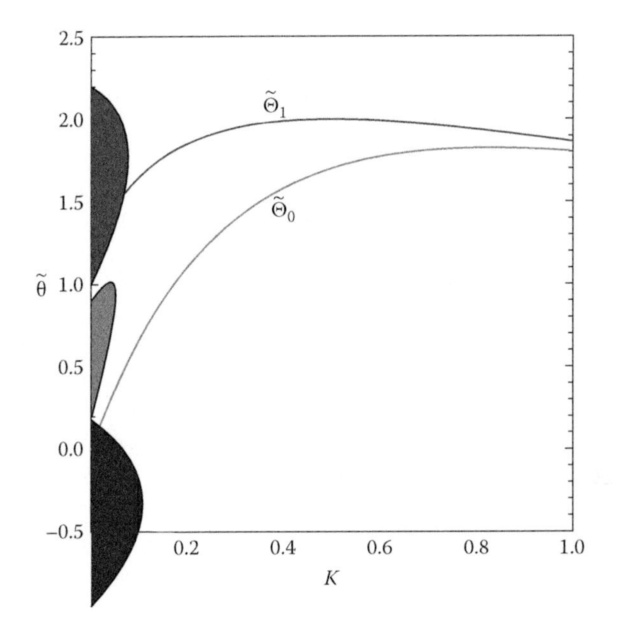

FIGURE 12.2 The phase diagram K versus $\tilde{\theta}$, where the rough regions of normality, cancer, and diabetes are colored in green, red, and blue, respectively. See color insert.

(i.e., the minimal concentration of the growth factor), such as the one indicated by the dash-dot-dot line in Figure 12.1. If a toggle switch has I_{off} smaller than I_0 (e.g., the black curve with $\alpha = 1$), then it is actually irreversible, although not by definition. To see why, consider the following scenario. Suppose initially the concentration of the growth factor is at the baseline ($I = I_0$) and AKT is not activated ($A = 0$). As I increases over the threshold I_{on}, the protein kinase AKT is activated ($A = 1$). Now let the growth factor decrease. Even if it reduces to the minimal level I_0, AKT is still activated, because I_{off}, being smaller than I_0, is not reachable.

In Figure 9.4, the curve $\tilde{\theta} = \tilde{\Theta}_1(K)$ separates the region of irreversible switches from the region of toggle switches. Because of the connection between irreversible switches and cancer, the region of cancer should be primarily above the curve. Because toggle switches with small positive I_{off} are also a cancer risk, as discussed above, the cancer region may slightly encroach upon the area of the toggle switch. The cancer region is colored in red in Figure 12.2.

12.2 TYPE 2 DIABETES

In Chapter 10, the fundamental decomposition of the PI3K-AKT-TOR pathway shed light on many things, including the orthogonal roles played by the parameters α and β. In this chapter, I show that aberrant α and β values may correspond to different mechanisms of type 2 diabetes, on the basis of which I classify type 2 diabetes into two subtypes: type 2α and type 2β. Common type 2 diabetes may be a combination of the two subtypes.

12.2.1 TYPE 2α DIABETES

Compared with the toggle switch, the monotone type of response is characterized by a stifled AKT activation and thus glucose intolerance. The blue curve ($\alpha = -3$) in Figure 12.1 is a typical example. If the pancreas produces a normal amount of insulin, then AKT activation is insufficient to drive the translocation of GLUT4 to the plasma membrane; muscular glucose uptake is impaired and glycemic control becomes difficult. To overcome this difficulty, the pancreas has to produce much more insulin so that AKT can be activated sufficiently to bring the plasma glucose concentration under control. That is, the plasma glucose level may still be normal, owing to insulin overproduction. But the response curve (which is more essential) has already been aberrant, probably for a long time. This asymptomatic state is known as *prediabetic*. The overproduction of insulin would gradually exhaust the pancreas (e.g., the destruction of islet cells) and lead to its failure. Finally, the pancreas can no longer secrete the amount of insulin it did previously. The weakened insulin secretion, together with the aberrant response curve, leads to full-blown diabetes: glucose cannot be utilized by the majority of tissue cells and has to remain in the blood (hyperglycemia). All in all, the deformation of bistability into the monotone type increases the risk of type 2 diabetes.

The region of monotone type is below the curve $\tilde{\theta} = \tilde{\Theta}_0(K)$ in Figure 12.2. Because K should be small, the diabetes phenotype should locate roughly in the blue region in Figure 12.2, which is characterized by $\tilde{\theta} < 0$. Because $\tilde{\theta} = \alpha\beta$ and β is always positive, this kind of diabetes can be characterized by $\alpha < 0$. I thus call it type 2α diabetes.

Because a toggle switch with overly small hysteresis (an overly small positive α) may also be inefficient for glucose uptake, the region of type 2α diabetes (colored in blue in Figure 12.2) may encroach slightly upon the area of the toggle switch.

Because $\alpha = \Phi - \epsilon\Psi$ and Φ is relatively constant, the condition $\alpha < 0$ is usually caused by an overly large Ψ (i.e., mTOR overactivity). It is well known that mTOR hyperactivity increases the risk of type 2 diabetes [34]. There are several immediate factors that can promote mTOR activation, including the pAKT level and the nutrient level. The pAKT level has been considered separately in the mathematical model; hence Ψ is not affected by the pAKT level. Therefore, the value of Ψ is affected primarily by the intracellular nutrient level. Among all the nutrients, amino acids are potent mTOR activators, especially the branched-chain amino acids. Increased protein consumption elevates circulating amino acids, which promotes mTOR hyperactivity (large Ψ) and may lead to a negative α and thus type 2α diabetes.

12.2.2 TYPE 2β DIABETES

This subtype of diabetes is caused by an overly small β. The fundamental decomposition tells us that the response curve can be translated along the I-axis without much distortion. Under the limit condition $K = 0$, one has

$$I_{\text{on}} = \gamma^{-1}\beta^{-1},$$
$$I_{\text{off}} = \gamma^{-1}\beta^{-1} - \gamma^{-1}\alpha.$$

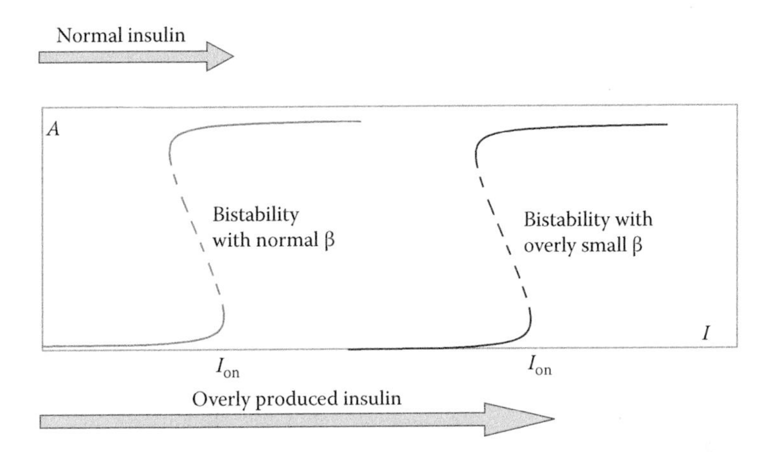

FIGURE 12.3 Insulin resistance caused by the decrease of β. The green curve corresponds to normal β, which allows for a normal amount of insulin for AKT activation. The blue curve corresponds to overly small β, which requires overproduction of insulin to activate AKT. See color insert.

Because γ is relatively constant, β is largely responsible for the shift in the response curve. As β decreases, both I_{on} and I_{off} are enlarged, and therefore the response curve shifts to the right. The shift does not involve any distortion; because $\Delta I = I_{on} - I_{off} = \gamma^{-1}\alpha$, the difference between I_{on} and I_{off} does not change with β. The green curve in Figure 12.3 illustrates the response curve of a cell with a normal β value. Because its I_{on} value is smaller than the peak of a normal insulin pulse, the cell can be normally activated to uptake glucose. The blue curve in Figure 12.3 illustrates a response curve with an overly small β and thus an overly large I_{on}, which cannot be reached by the normal insulin pulse. In this event, the pancreas has to overproduce insulin so that the peak insulin level can reach I_{on}. The person is in an asymptomatic, prediabetic state. The plasma glucose level can still be controlled within the normal range but the control depends on the heavily overloaded pancreas. The prolonged insulin overproduction gradually damages the pancreas and impairs its insulin secretion capacity. Finally, full-blown diabetes and glucose intolerance ensue. Figure 12.3 implies that type 2 diabetes is not exclusively caused by the monotone type; a bistable response with an overly small β (an overly large I_{on}) can also induce insulin resistance and type 2 diabetes. I thus call it type 2β diabetes.

Although an abrupt reduction of β can happen due to acute gene mutations, a more common scenario might be that the β value is normal initially but decreases gradually over time. For instance, the gradual decline of β can be induced by the gradual accumulation of fat in the body. As an efficient fuel, fat is important for the normal functioning of the body. But excessive fat causes many health problems. In the body, adipocytes are specialized to store fat by sequestering lipids in droplets. Although adipocytes are highly capable to do that, their capacity is not without limitations. An overweight person has such a high lipid content that the esterification of lipids becomes increasingly difficult as the person becomes fatter. As a consequence, the FFA level becomes elevated as a person becomes obese. This increases the diglyceride (DAG) level, which in turn activates protein kinase C theta (PKC-θ), a serine kinase that causes increased serine phosphorylation of IRS. Such serine phosphorylation would reduce the ability of IRS to activate PI3K [13]. As a consequence, the β value is reduced and the I_{on} value is increased. The response curve $A(I)$ of an average adipocyte of an obese person thus shifts toward the right, when compared with that of a normal person. That is, there is a systematic (i.e., applicable to most adipocytes in the body) elevation of the I_{on} value when a person becomes obese.

The ill effects of excessive fat extend to cells other than adipocytes, particularly myocytes. As *free* fatty acid, FFA can diffuse into muscles and increase the content of intramuscular triglyceride (IMTG), an important energy source for muscle contraction during exercise. The content of IMTG is determined by many factors, including diet, type of muscle fiber, and gender [18]. Among them diet is an obvious factor. A high-fat diet not only exacerbates the state of obesity, but also increases the concentration of IMTG. Indeed, LCFAs from circulating albumin-bound fatty acids and from VLDL-TGs are too many to be contained by adipocytes; some of them inevitably deposit in ectopic places such as muscles. Note that fat deposition in skeletal muscles is facilitated by LPL, which actively degrades LCFAs. Unfortunately, an increased amount of fat in the diet enhances LPL activity. The increased IMTG concentration promotes the activation of PKC-θ, which interferes with IRS-PI3K interaction in the myocytes and reduces their β values. This causes a shift to the right of the response curves $A(I)$ of myocytes.

In summary, obesity and/or a high-fat diet would cause systematic right-shifts of the response curves of both adipocytes and myocytes. In other words, both adipocytes and myocytes become insulin resistant, due to obesity. Figure 12.4 illustrates the shift phenomena for both adipocytes and myocytes, as a person becomes obese. Heterogeneity is also considered among the cells in terms of their I_{on} values. For example, a larger adipocyte might have a larger I_{on} value than a smaller adipocyte, because it contains more lipids. As such, a myocyte having more (less) IMTG might have a larger (smaller) I_{on} value. In Figure 12.4, the red curve represents the number distribution of myocytes over their I_{on} values; the black curve represents the number distribution of adipocytes over their I_{on} values. As the person becomes obese, both distributions drift to the right, probably in a synchronized manner. This necessitates overproduction of insulin by the pancreas to overcome the globally increased I_{on}. This prediction is in line with clinical data (Figure 12.5) obtained from 14 normal and 15 obese persons. One sees clearly that the plasma insulin concentrations of obese patients are much higher than those of normal controls.

12.3 PATHOLOGIC CONDITIONS CAUSED BY AN ENLARGED K

The above discussions focused on the normal condition that K is small. One can even consider the limit condition $K \to 0$ only, which approximates well the condition that K is positive and small. Indeed, the effect of $K \to 0$ is just to turn the smooth curves (see Figure 10.2(b)) into straight lines (Figure 10.2(a)), whereas the key features of the curves do not change.

When K is overly large, the response curves become "flat" — insensitive to the change in insulin concentration (Figure 10.2(c)). At low insulin concentrations, the response is fairly high (although it is desired to be nearly zero). At high insulin concentrations, the response is not high enough (although it is desired to be nearly full).

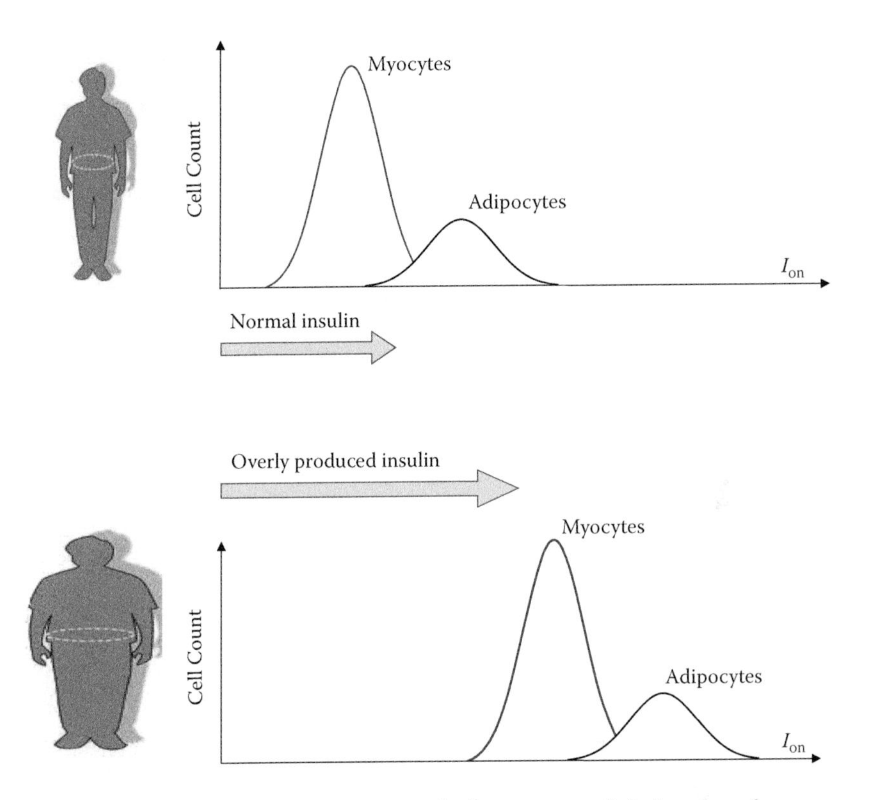

FIGURE 12.4 The distribution of myocytes and adipocytes over their I_{on} values, for a person in both the lean (upper panel) and obese (lower panel) states. As the person becomes obese, the I_{on} values shift to the right, for both myocytes and adipocytes. The pancreas has to secrete more insulin to trigger glucose uptake. See color insert.

Because sensitivity is of fundamental importance in biology, insensitivity caused by a large K would have global ill effects. In particular, it increases the risk of both diabetes and cancer.

The diabetes risk is due to insufficient AKT activation even at very high insulin concentrations. When K is small, the full activation of AKT ($A \approx 1$) is easy. For the case $K = 0.01$ and $\alpha = 0.6$ (the green curve in Figure 10.2(b)), AKT is fully activated at $I = 1$. When K is large, the full activation of AKT becomes difficult. For the case $K = 1.00$ and $\alpha = 0.6$ (the green curve in Figure 10.2(c)), AKT is only activated in half ($A \approx 0.6$) at $I = 1$. Theoretically speaking, it is possible to make A approach one for the case $K = 1.00$, but that demands a nearly infinite amount of insulin stimulation.

Note that A is only a relative value, namely, the percentage of activated AKT. The reduction of absolute AKT activation (i.e., [pAKT]) caused by an increased K is even more dramatic. Because

$$K = K_m / A_{\max} \tag{12.1}$$

and the Michaelis constant K_m is relatively constant, the increase of K is actually caused by the decrease of $A_{\max}$, namely, the total AKT concentration. In other words,

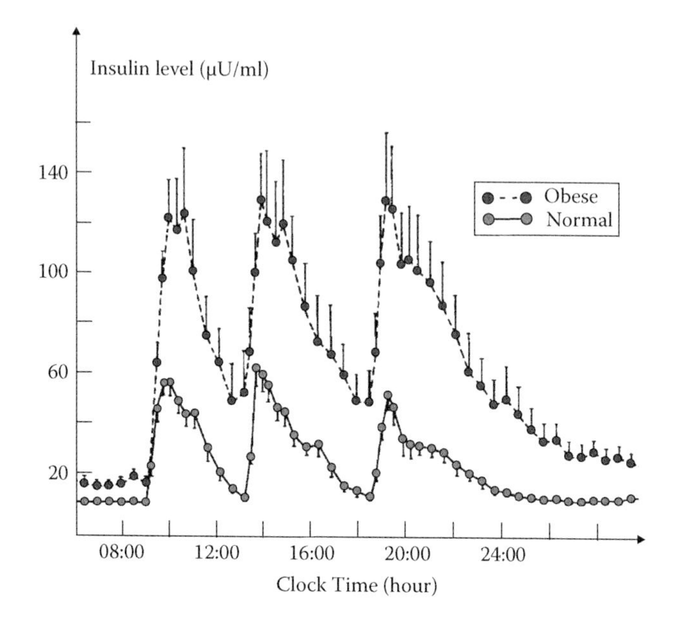

FIGURE 12.5 Twenty-four-hour profile of plasma insulin concentration averaged from 14 normal (circles with solid lines) and 15 obese (circles with dashed lines) subjects. There are three spikes, caused by the three meals at 09.00, 13.00, and 18.00. (Adapted from [24].)

the increase of K reduces *both* the total AKT ($A_{\max}$) and the percentage of activated AKT (A). As a consequence, the absolute AKT activation is dramatically reduced, because

$$[\text{pAKT}] = A \cdot A_{\max}. \tag{12.2}$$

This diminished capability of AKT activation would impair glucose uptake by tissue cells (primarily in muscles and fat), exacerbate glucose intolerance, and increase the risk of diabetes.

The increase of K also increases cancer risk, primarily due to the positive correlation between diabetes and cancer (see below).

12.4 RELATIONSHIP BETWEEN CANCER AND DIABETES

Figure 12.2 seems to suggest that cancer and diabetes are opposite to each other. Does that mean cancer and diabetes are mutually exclusive? That is, does a cancer patient have less chance to contract diabetes? And does a diabetes patient have less chance to contract cancer? The answer is no. Clinical and epidemiological studies have established the positive correlation between cancer and metabolic syndrome. Patients diagnosed with metabolic diseases show an increased incidence and aggressiveness of cancer development. Conversely, patients with diabetes treated with metformin to lower insulin levels have reduced chances of cancer in comparison with untreated individuals.

The reason is that we are multicellular organisms. Different cell types, although having the same PI3K-AKT-TOR pathway structure, may respond differentially to insulin or other growth factors. Even for the same cell, its response curve may be different under different metabolic conditions. It is thus possible that some cells are insulin resistant (e.g., with a negative α or a very small β); some cells are insulin hyperresponsive (with $\tilde{\theta}$ greater than or close to 1); and some cells are just normal. If, for example, skeletal muscles become insulin resistant, then the person would be prediabetic or diabetic. This state does not prevent a cell of another type from becoming cancerous. Therefore, cancer and diabetes can in principle coexist in the same person. They are not mutually exclusive at all.

Then, how to explain the positive correlation between the two diseases? The major cell type that becomes resistant to insulin during the pathogenesis of diabetes is the myocyte. Because muscles make up a large portion of our body weight, their resistance to insulin leaves a high concentration of glucose in the blood. If muscles made up a small portion of our body weight, then muscles' insulin resistance would only perturb slightly whole-body glucose homeostasis. Because of cancer cells' glucose addiction (cancer's sweet tooth), hyperglycemia certainly provides a favorable environment for cancer growth and even exerts a selection pressure for carcinogenesis. On the other hand, hyperinsulinemia, a hallmark of obesity and diabetes, also favors cancer growth because insulin acts as a potent mitogen. Diabetes would certainly increase cancer risk.

12.5 IMPLICATIONS FOR THERAPY AND DISEASE PREVENTION

Our analyses have revealed the importance of the three parameters K, α, and β in controlling the pathway's functional output and the correspondence between abnormal changes of the three parameters and various pathologic phenotypes. Therefore, therapeutic interventions can potentially be realized through modulating the three parameters, so that aberrant responses can be adjusted back to normal. An efficacious therapy consists in a minimal-effort perturbation of the parameters back to their normal regions (e.g., from the red region to the green region in Figure 12.2; or from the blue region to the green region).

12.5.1 THE PARAMETER K

Because K is inversely proportional to $A_{\max}$, the modulation of K is actually realized through targeting $A_{\max}$, the total AKT concentration. Because AKT is an oncoprotein, one would expect that a higher concentration of total AKT leads to a higher level of pAKT and thus a higher chance of cancer. One would also expect that inhibition of total AKT (through suppressing AKT gene expression) is a sound strategy to inhibit the growth of cancer cells. In fact, there have been many experimental studies and clinical trials on inhibiting cancer growth through suppressing AKT gene expression (e.g., [38]). Our analysis has raised an alarm about these activities, because the abundance of total AKT is a good thing as far as dynamical properties are concerned. It causes a small value of K and thus the sensitivity of response to insulin or other growth factors that is important in maintaining a person's well-being. Any therapeutic strategy involving the inhibition of AKT gene expression would have negative effects and should be used with caution.

In fact, it is the duration of AKT activation that is critical to carcinogenesis, not the peak AKT activation. Constitutive elevation of pAKT implies continuous and tonic glucose uptake, which exerts a selection pressure favoring malignant transformation. A large peak pAKT value (i.e., total AKT concentration) is perfectly fine, as long as the response curve maintains good shape (a sensitive toggle switch with suitable I_{on} and I_{off} values). The inhibition of total AKT would increase K and thus *dramatically* reduce $[\mathrm{pAKT}] = A \cdot A_{\max}$, because $A_{\max}$ is reduced (a direct effect of inhibition) and A is also reduced (an indirect effect through the PI3K-AKT-TOR pathway). For a systemic drug (i.e., a drug that affects the body as a whole) inhibiting total AKT, the consequential insulin resistance may be more severe than one expects. That is, the inhibition of total AKT would increase the risk of diabetes. Because of the positive correlation between diabetes and cancer, the inhibition of total AKT, although aimed to suppress cancer, paradoxically favors cancer development.

Total AKT abundance is only a necessary condition for normal AKT activation. Aberrations of other molecules also disturb the system-level control mechanism, the best known of which include the aberrations of TOR activity and PI3K activity. In the following, it will be shown that the two aberrations are orthogonal to each other and can be recapitulated as the perturbations of the two parameters α and β.

12.5.2 THE PARAMETER α

TOR is a popular drug target for cancer therapy. The inhibition of TOR slows macromolecular synthesis and thus cancer cell growth. Despite the sound rationale, the method has had only modest success in clinical trials and the results were often unpredictable. The equation

$$\tilde{\theta} = \alpha\beta = (\Phi - \epsilon\Psi)\gamma A_{\max}\beta \tag{12.3}$$

is a reproduction of Equation (11.1), which can explain why TOR inhibition is not effective: it reduces Ψ but increases $\tilde{\theta}$, which deforms $A(\tilde{I})$ close to or into the cancer region (colored in red in Figure 12.2). This theoretical result agrees with the clinical observation that TOR inhibition actually induces upstream receptor tyrosine kinase signaling and activates AKT. In brief, although TOR has some aspects beneficial to cancer growth, it also has aspects that are adverse to cancer growth. Therefore, the inhibition of TOR also has dual effects and thus the clinical outcome is unpredictable.

12.5.3 THE PARAMETER β

According to Equation (12.3), the inhibition of TOR (i.e., the reduction of Ψ) for cancer therapy would cause the increase of $\tilde{\theta}$, which is paradoxically favorable to cancer growth. To counteract this undesirable effect, the other parameters in Equation (12.3) should be perturbed so that a normal $\tilde{\theta}$ can be maintained. Because Φ, ϵ, and γ are relatively constant, the target of perturbation should be $A_{\max}$ and β. The parameter $A_{\max}$ can be reduced by inhibiting the total AKT, which is not recommended, as demonstrated in the above discussions. The parameter β can be reduced by inhibiting PI3K, which is recommended for future clinical trials.

12.5.4 Combinatorial Therapy

Compared with traditional therapies that have less directed modes of action (e.g., chemo- and radiotherapy), molecule-targeted therapy represents a great advancement in therapeutics because it bestows a potentially more reliable performance while significantly reducing toxicity to patients. However, molecules in a cell do not function in isolation; it is the entire network of interactions between the molecules that determines system-level control mechanisms. Indiscriminately targeting individual molecules in a biomolecular network, without an understanding of the control mechanisms in which they participate, could give rise to unexpected and highly undesirable results. It is thus important to reveal the control mechanisms through sophisticated mathematical modeling and analysis, which also allow us to derive strategies to modulate the control mechanisms. Our hope of conquering complex diseases may consist in a smart modulation of the control mechanisms in favor of the return to the healthy state.

Our analysis of the PI3K-AKT-TOR pathway reveals that targeting a single molecule usually involves undesired effects. The inhibition of total AKT would degrade the fine dynamical properties of the pathway (sensitivity, adaptivity, etc.) and would exacerbate cancer, diabetes, and possibly other diseases. The inhibition of TOR would disrupt the negative feedback loop in the pathway, which paradoxically increases cancer risks. It is thus necessary to target multiple molecules, by using a drug cocktail consisting of an array of specific agents. Further mathematical analysis, together with experimental studies, may figure out the appropriate proportions of drug agents so as to obtain the maximum synergistic effect. This will provide direct guidance for producing a low-dosage drug with minimal toxicity and drug resistance.

13 Tao of Diseases

The past century witnessed the simultaneous increase of many complex diseases, which continues to escalate into this century. In the 1950s to 1960s the incidence of diabetes was about 2% and today it is about 8%. The escalation of diabetes has been paralleled by that of obesity and cancer, and some other complex diseases as well. These epidemiological data demonstrate a deep connection among the diseases. Because the epidemics escalate at the *same* time, the diseases must have a common molecular basis, such as involving the same set of genes, or having a similar network of genetic regulation. On the other hand, these epidemiological data also reveal the evolutionary underpinning of the diseases. Because the epidemics escalate in *our* time, environmental and lifestyle changes should play an important role in the etiology and pathogenesis of complex diseases. It is thus clear that complex diseases involve the intricate interplay among molecular, physiological, environmental, lifestyle, and possibly other factors, contributing to their epidemics and exacerbating their treatments.

Complex diseases are difficult to cure. Despite tremendous technological developments which allow our observations and treatments to penetrate into the molecular level, the overall rate of cancer mortality has been basically flat for a long time, and diabetes remains an incurable malady. Chemotherapy and radiotherapy can kill a subset of cancer cells, but the remaining cancer cells soon develop drug-resistant potency. Molecule-targeted therapy can potentially bestow more reliable performance while significantly reducing toxicity to patients. Unfortunately, the usefulness of the approach is limited and patients often relapse from the disease. A complex disease thus behaves like a master of martial arts who has an extraordinary capacity to evade attack and fight back. It has obtained "enlightenment" for survival and development. In the language of traditional Chinese philosophy, the disease has obtained *Tao*. As a manifestation of the Tao, every treatment of the disease would create counteractions to neutralize the treatment. First, the treatment would often lead to unpredictable and undesirable side effects, which limit further treatments. Second, the disease would remain robust in response to the treatment, possibly even with enhanced malignancy.

The first aspect has its roots in biocomplexity. Indeed, the complexity of a disease is largely inherited from the complexity of molecular interactions. In this book we have learned principles for organizing a myriad of biomolecules into a network of interactions, which include structural properties such as modularity, recurring motifs, and scale-free connections; and dynamical properties such as bistability, sensitivity, and adaptivity. We have also seen many Yin-Yang pairs, such as the phosphorylation and dephosphorylation of a protein kinase, and positive and negative feedback. Because every molecule is connected with others in a rather complex manner, targeting one molecule inevitably affects other molecules and produces unexpected results (e.g., the disruption of feedback loops, the reduction of sensitivity, etc.). In other words, it is difficult to find real targets because everyone seems to be a target.

The second aspect implies that complex diseases may correspond to well-evolved, optimized control mechanisms that are useful or even indispensable under certain (possibly harsh) conditions. These extreme conditions may have been common during certain periods in evolution but are largely removed in modern society. As such, some control mechanisms may be required during certain stages of cell development but are not required any more upon maturation; or they may be required for certain types of cell but not for others. These control mechanisms have to be very robust, given that they have to function well under extreme conditions. Therefore, the robustness of the control mechanisms might be a fine property acquired through evolution, which was necessary for the survival of the organism. If the mechanisms are activated in the wrong cell or at the wrong time, however, diseases may occur. And the diseases are difficult to cure because they are powered by rather robust and complex control mechanisms.

In this book, the molecular bases of complex diseases were considered in the context of the PI3K-AKT-TOR pathway. Although this seems to be an ad hoc choice, its study actually has general significance. First, the PI3K-AKT-TOR pathway is a central circuit controlling glucose metabolism, cell growth, proliferation, and survival, among many important biological processes. The deregulation of the pathway is highly relevant to metabolic diseases, including cancer and diabetes. Second, the PI3K-AKT-TOR pathway is evolutionarily well conserved. It presents in almost every cell of every multicellular organism. Its backbone structure is essentially the same in human cells as in the cells of fruit flies or worms, although the homologue proteins may have different names. The conclusions in this book, being based on the network structure and being independent of the parameters, are quite general. Finally, we have proved that bistability is a universal mechanism for optimal homeostatic control. Some of our results, although derived in the context of the PI3K-AKT-TOR pathway, are applicable to other pathways that use positive feedback to realize multistability.

Analysis of the PI3K-AKT-TOR pathway has revealed that there are only three kinds of control mechanisms in the form of response curves: the monotone type, the toggle switch, and the irreversible switch. It is clear that the toggle switch should be a normal mode of regulation, but this does not rule out the possibility that the monotone type and irreversible switch are also used by some species or in certain cell types. Indeed, there are innumerable multicellular species, and different species live in disparate environments and have disparate physiology. It is thus unlikely that a single toggle switch can handle the whole spectrum of metabolic conditions. That is, the monotone type and irreversible switch may be adopted by some species whose physiology is quite different from ours. Even for the same organism, ever-changing metabolic conditions may also drive the deformation of the response curve, and it is possible that a qualitative transition occurs temporarily in order to cope with an extremely harsh condition. Furthermore, even for the same organism at a single time, it is possible that more than one kind of control mechanism may coexist, due to a wide variety of cell types in the body. Therefore, the mechanisms should all be robust so that they can withstand extensive molecular perturbations within the cell, especially those mechanisms streamlined for extreme conditions. In the following, diabetes and cancer are proposed to be driven by well-evolved mechanisms.

Diabetes corresponds to insulin resistance, which has long been known as a harmful state. The identification of bistability (especially the threshold I_{on}) implies that insulin resistance is an indispensable mechanism to guarantee the brain's glucose supply. Because *two heads are more expensive than one*, insulin resistance is even more crucial and should be greater for a woman during pregnancy. That is, the I_{on} value of the mother's muscles should be increased to allow for more glucose to be shunted away from the mother's muscles and to be diverted to the fetal brain in addition to the mother's brain. This mechanism is crucial to cope with food scarcity, which was quite common in the past. After the invention of agriculture and industry, the dark side of insulin resistance emerged and becomes dramatic (in the form of diabetes) in modern society that is characterized by overnutrition and physical inactivity. In this book, we have learned how a molecular mechanism (insulin resistance) intertwines with an environmental factor (overnutrition) and a lifestyle factor (physical inactivity). The molecular mechanism itself is nice, but it can turn into good or evil depending on other factors.

Cancer corresponds to constitutive activation of oncoproteins. In the PI3K-AKT-TOR pathway, the constitutive activation is rendered by the irreversible phosphorylation of AKT when the growth factor level exceeds I_{on}. The irreversible switch is characterized by a negative threshold I_{off}, which is unreachable because the growth factor level must be positive. Once the oncoprotein is activated, it is always activated, even after a complete withdrawal of growth factor stimulation. Although it may cause cancer when used ectopically, the irreversible switch proves to be an indispensable control mechanism for several important biological processes. For instance, cells progressing through the cell cycle must commit irreversibly to mitosis without slipping back to interphase before properly segregating their chromosomes. It has been demonstrated in *Xenopus laevis* egg extracts that cell cycle transitions into and out of mitosis are controlled by an irreversible switch [28]. Cell maturation is also a one-way process, in which irreversibility is required for commitment to differentiation. It was demonstrated that the maturation of *Xenopus* oocytes is controlled by an irreversible switch, which produces an actively maintained "memory" of a transient inductive stimulus and could explain the irreversibility of maturation [37].

The Tao of a general complex disease can be similarly discussed. It involves the intricate interplay between fine control mechanisms and disease causative agents. The latter is broadly defined to include pathogens, toxic chemicals, adverse environmental and lifestyle factors, and so on. Just like a virus that can hijack the host's resources to sustain its replication and to evade immunological destruction, a disease causative agent can take advantage of the host cell's control mechanisms to power disease development and to resist treatments. The control mechanisms were once indispensable for survival, but are usually not operative in a healthy, present-day person, or are operative only in certain cells or during certain processes. If they are activated by disease causative agents at the wrong time or in the wrong cell, aberrations occur. Here I use the word "aberrations" simply because I know their future pathological consequences. But they are quite normal to the cell, because the ectopic control mechanism that underlies the aberrations is a long-lost friend of the cell. Therefore, the aberrations are not detected by the immune system, which allows for their asymptomatic and smooth augmentation. In this sense, the control mechanism is robust and can persistently drive disease progression throughout the latency phase and continue into the full-blown phase of the disease. The robustness further compromises therapeutic interventions. In spite of great scientific and technological advancement and vast amounts of monies spent, we are still nowhere close to a satisfactory solution.

This book has provided a framework of global analysis that may be one important step toward solving our health care crises. Through the elucidation of the Tao of diseases, we now know that the global picture we look at should be called a picture of control mechanisms, instead of diseases, to be more precise. One important contribution of this book is the geometrization of control mechanisms into response curves, which makes rigorous and insightful mathematical analysis possible. The identification of control mechanisms (classification of response curves) allows for decompositions of both the network structure and the biological functions. Such decompositions allow for a one-to-one mapping between disease causative factors and their effects on the control mechanism. For example, we found that type 2 diabetes may be further divided into two categories: type 2α and type 2β. A high-fat diet would primarily shift the bistable switch with no or little distortion and would increase the risk of type 2β diabetes, whereas a diet of high amino acids would primarily twist the bistable switch with no or little shift and would increase the risk of type 2α diabetes. Because an obese person usually has elevation of both fat and amino acid levels, type 2 diabetes may well be a combination of type 2α and type 2β. We expect the framework will provide more insights into the intricate interplay among the multiple factors affecting complex diseases and will contribute to disease prevention and therapies.

A Optimization of Insulin Action

A.1 PROBLEM FORMATION

The objective of the optimization is to find the control $u(t)$ that restores homeostasis in the shortest time, with the constraint that the average of the control is fixed at a small value $\eta \cdot u_{\max}$, where $u_{\max}$ is the upper bound of the control and η is small. That is,

$$\min_{u(t)} T \quad \text{subject to}$$
$$\frac{1}{T}\int_0^T u(t)dt = \eta u_{\max}. \tag{A.1}$$

To incorporate this constraint into the analysis, a new state variable $W(t)$ is introduced that satisfies

$$\frac{dW}{dt} = u - \eta \cdot u_{\max} \tag{A.2}$$

with the condition $W(0) = W(T) = 0$. Note that Equation (A.2) is equivalent to Equation (A.1). Indeed, the integration of Equation (A.2) yields $W(T) - W(0) = \int_0^T u(t)dt - \eta u_{\max} T$, which is exactly Equation (A.1) by the condition $W(0) = W(T) = 0$.

Therefore, the "constrained" glucose-insulin feedback system is three dimensional.

$$\frac{dG}{dt} = s - \lambda G - u(t)G + m(t), \tag{A.3}$$
$$\frac{dI}{dt} = f(G) - kI, \tag{A.4}$$
$$\frac{dW}{dt} = u - \eta u_{\max}. \tag{A.5}$$

It is a nonautonomous system, because the function $m(t)$ depends explicitly on t. The system can be written in the vector form

$$\dot{\mathbf{X}} = \mathbf{F}(\mathbf{X}, u, t), \tag{A.6}$$

where

$$\mathbf{X} = [G, I, W]$$

and

$$\mathbf{F} = [F_G, F_I, F_W]$$

are vectors of functions representing the right-hand sides of Equations (A.3–A.5). Starting from the initial state $X(0) = [G_0, I_0, 0]$, the system evolves and reaches the final state $X(T) = [G_0, I_0, 0]$, i.e., the system returns to where it starts, $X(T) = X(0)$, after time T.

Pontryagin's maximum principle introduces a vector of Lagrange multipliers that is adjoint to the state vector $\mathbf{X}$. The Lagrange multipliers are denoted by lower case letters of their counterparts: $\mathbf{x} = [g, i, w]$. They together constitute the Hamiltonian system:

$$\frac{dG}{dt} = \frac{\partial H}{\partial g}, \tag{A.7}$$

$$\frac{dI}{dt} = \frac{\partial H}{\partial i}, \tag{A.8}$$

$$\frac{dW}{dt} = \frac{\partial H}{\partial w}, \tag{A.9}$$

$$\frac{dg}{dt} = -\frac{\partial H}{\partial G}, \tag{A.10}$$

$$\frac{di}{dt} = -\frac{\partial H}{\partial I}, \tag{A.11}$$

$$\frac{dw}{dt} = -\frac{\partial H}{\partial W}. \tag{A.12}$$

Here

$$H(\mathbf{x}, \mathbf{X}, t, u) = gF_G + iF_I + wF_W \tag{A.13}$$

is the Hamiltonian function. Equations (A.7–A.9) reiterate Equations (A.3–A.5). Equations (A.10–A.12) can be extended into the following:

$$\frac{dg}{dt} = (\lambda + u)g - f'i, \tag{A.14}$$

$$\frac{di}{dt} = ki, \tag{A.15}$$

$$\frac{dw}{dt} = 0, \tag{A.16}$$

where f' denotes df/dG, the differentiation of $f(G)$ with respect to G. From the equations one obtains

$$i(t) = i_0 e^{kt}$$

and

$$w(t) = w_0,$$

where $i_0 = i(0)$ and $w_0 = w(0)$ are initial values of $i(t)$ and $w(t)$.

The general time-optimal problem for a nonautonomous system was solved previously. In the context of the present problem, the solution can be described as the following theorem.

Theorem A.1 *Let $u(t)$, $0 \le t \le T$, be a control that transfers the system described by Equations (A.3–A.5) from the state $\mathbf{X}(0)$ to the state $\mathbf{X}(T)$. Let $\mathbf{X}(t)$ be the corresponding trajectory. The control $u(t)$ is locally optimal (in terms of minimizing T),* *if and only if there exists a nonzero continuous vector function* $\mathbf{x}(t)$ *(satisfying Equations (A.14–A.16)), such that:*

(i) For all t, $0 \le t \le T$, *the Hamiltonian function* $H(\mathbf{x}(t), \mathbf{X}(t), t, v)$ *of the variable* $v \in [0, u_{\max}]$ *attains its maximum at the point* $u(t)$. *That is,*

$$H(\mathbf{x}(t), \mathbf{X}(t), t, v) \le H(\mathbf{x}(t), \mathbf{X}(t), t, u(t))$$

for all v *in the range* $0 \le v \le u_{\max}$.

(ii) The relation

$$H(\mathbf{x}(T), \mathbf{X}(T), T, u(T)) \ge 0$$

is satisfied.

Note that conditions (i) and (ii) do not guarantee that $u(t)$ achieves the smallest T (the global optimum). Therefore, conditions (i) and (ii) are necessary, but not sufficient, for global optimality.

A.2 EMERGENCE OF THE ALL-OR-NONE PROPERTY

From Equation (A.13) one obtains the expression of the Hamiltonian

$$\begin{aligned} H(\mathbf{x}, \mathbf{X}, t, u(t)) &= g(s - \lambda G - u(t) G + m(t)) \\ &\quad + i(f(G) - kI) + w(u(t) - \eta u_{\max}) \\ &= \rho u(t) + g(s - \lambda G + m(t)) \\ &\quad + i(f(G) - kI) - \eta u_{\max} w, \end{aligned} \tag{A.17}$$

where

$$\rho(t) = w(t) - g(t) G(t). \tag{A.18}$$

To maximize the Hamiltonian as required by condition (i) of Theorem A.1, one needs to set $u(t) = u_{\max}$ whenever $\rho(t) > 0$ and $u(t) = 0$ whenever $\rho(t) \le 0$. That is

$$u(t) = \begin{cases} 0 & \rho(t) \le 0 \\ u_{\max} & \rho(t) > 0. \end{cases} \tag{A.19}$$

The function $\rho(t)$ is called the switching function because it dictates the switching behavior of the optimal control $u(t)$.

To fulfill condition (ii) of Theorem A.1, the Hamiltonian function at $t = T$ is computed:

$$\begin{aligned} H(\mathbf{x}(T), \mathbf{X}(T), T, u(T)) &= \rho(T) u(T) + g(T)(s - \lambda G(T) + m(T)) \\ &\quad + i(T)(f(G(T)) - kI(T)) - \eta u_{\max} w(T). \end{aligned} \tag{A.20}$$

Because at time T the system is in the fasting state, one has $u(T) = 0$, $m(T) = 0$, $G(T) = G_0$, and $I(T) = I_0$. Since the fasting state is also the steady state, one has $s - \lambda G_0 = 0$ and $f(G_0) - kI_0 = 0$. One therefore has

$$H(\mathbf{x}(T), \mathbf{X}(T), T, u(T)) = -\eta u_{\max} w_0. \tag{A.21}$$

Therefore, condition (ii) implies that $w_0 \leq 0$ must hold in order that $u(t)$ is optimal. This requirement will be fulfilled later as a part of constraining $\rho(t)$.

Although $\rho(t)$ has not been determined yet, the analysis thus far has already obtained significant conclusions. To be optimal, $u(t)$ cannot be a continuous, graded function. Instead, it is a discontinuous, piecewise constant function; it switches between the minimum and the maximum, but never takes any values in between.

A.3 SWITCHING FUNCTION $\rho(T)$

To obtain the switching function $\rho(t)$, one first obtains the differential equation governing the evolution of $\rho(t)$:

$$\frac{d\rho}{dt} = \frac{d(w - gG)}{dt} = \frac{dw}{dt} - \frac{dg}{dt}G - g\frac{dG}{dt}. \tag{A.22}$$

By combining Equation (A.22) with Equations (A.3, A.14, and A.16), one obtains

$$\frac{d\rho}{dt} = p(t)\rho + q(t) \tag{A.23}$$

where

$$p(t) = (s + m)/G$$

and

$$q(t) = i_0 f' G e^{kt} - w_0 p(t).$$

It is a first-order linear differential equation with variable coefficients, which has solutions in the closed form:

$$\rho(t) = e^{A(t)}(\rho_0 + i_0 B(t) - w_0 C(t)), \tag{A.24}$$

where $\rho_0 = \rho(0)$ is the initial value of $\rho(t)$,

$$A(t) = \int_0^t p(\tau)\, d\tau, \tag{A.25}$$

$$B(t) = \int_0^t G(\tau) f'(G(\tau)) e^{k\tau - A(\tau)} d\tau, \tag{A.26}$$

$$C(t) = \int_0^t p(\tau) e^{-A(\tau)} d\tau. \tag{A.27}$$

The functions $A(t)$, $B(t)$, and $C(t)$ are all monotonically increasing, because their kernels of integration are all positive. Note that $f'(G)$ is positive because $f(G)$ is an increasing function. Let

$$D(t) = B(t)/C(t) \tag{A.28}$$

for later use.

The function $\rho(t)$ depends on both the system parameters and the parameters ρ_0, i_0, and w_0, namely, the initial values of the functions $\rho(t)$, $i(t)$, and $w(t)$, respectively. By specifying the three parameters, a $\rho(t)$ and the corresponding $u(t)$

can be constructed pointwise along the t-axis. The following narrative explains the transitions $t \to t + \Delta t \to t + 2\Delta t$, where Δt is the step size.

Given the values of $G(t)$, $\rho(t)$, and $u(t)$ at time t, one first integrates the differential equation (A.29) to obtain $G(t + \Delta t)$. One then uses the obtained $G(t + \Delta t)$ to calculate $A(t + \Delta t)$, $B(t + \Delta t)$, and $C(t + \Delta t)$. Subsequently, the value of $\rho(t + \Delta t)$ is calculated by Equation (A.30). Depending on the sign of $\rho(t + \Delta t)$, the value of $u(t + \Delta t)$ is determined by Equation (A.31). Now that $G(t + \Delta t)$, $\rho(t + \Delta t)$, and $u(t + \Delta t)$ are all obtained, the above process is repeated to obtain $G(t + 2\Delta t)$, $\rho(t + 2\Delta t)$, and $u(t + 2\Delta t)$.

$$\frac{dG(t)}{dt} = s - (\lambda + u(t))\, G(t) + m(t), \tag{A.29}$$

$$\rho(t) = e^{A(t)} (\rho_0 + i_0 B(t) - w_0 C(t)), \tag{A.30}$$

$$u(t) = \begin{cases} 0 & \rho(t) \le 0 \\ u_{\max} & \rho(t) > 0. \end{cases} \tag{A.31}$$

A.4 OPTIMAL $U(T)$ AS A SQUARE PULSE

From the above we know that the optimal $u(t)$ is an all-or-none function that switches between 0 and $u_{\max}$. Because of the constraint $u(0) = u(T) = 0$, the function $u(t)$ must switch an even number of times $N = 2, 4, 6, \cdots, \infty$. Indeed, if $u(t)$ switches an odd number of times, then one must have $u(0) \ne u(T)$, which violates the constraint. As a matter of fact, an optimal control can only switch twice ($N = 2$). In the following, I will show that a control with $N = 4$ cannot be optimal. The conclusion can be readily extended to the case $N = 6, \cdots, \infty$.

Let's begin with the following question: given four different time points $0 < t_1 < t_2 < t_3 < t_4$, is the corresponding $u(t)$ (see Figure A.1(a)) an optimal control? To prove it is, one needs to find a $\rho(t)$ traversing the t-axis at t_1, t_2, t_3, and t_4. One then needs to make sure that the optimality is not singular.

The zeroes of the equation $\rho(t) = 0$, namely,

$$\rho_0 + i_0 B(t) - w_0 C(t) = 0, \tag{A.32}$$

correspond to the time points at which $\rho(t)$ traverses the t-axis. The discussion of Equation (A.32) is divided according to the sign of ρ_0.

- $\rho_0 > 0$. This condition implies an optimal control with $u(0) = u_{\max}$ (in order to maximize the Hamiltonian at time $t = 0$; see Equation (A.17)). However, $u(t)$ must satisfy $u(0) = 0$. Therefore, the function $u(t)$ cannot be an optimal control and the choice $\rho_0 > 0$ is not suitable.
- $\rho_0 = 0$. By Equation (A.32) one has

$$i_0 B(t_i) - w_0 C(t_i) = 0$$

for $i = 1, 2, 3, 4$, which represents four straight lines passing through the origin of the i_0 versus w_0 plane with the slopes $D(t_i)$ (Figure A.1(b)). Since

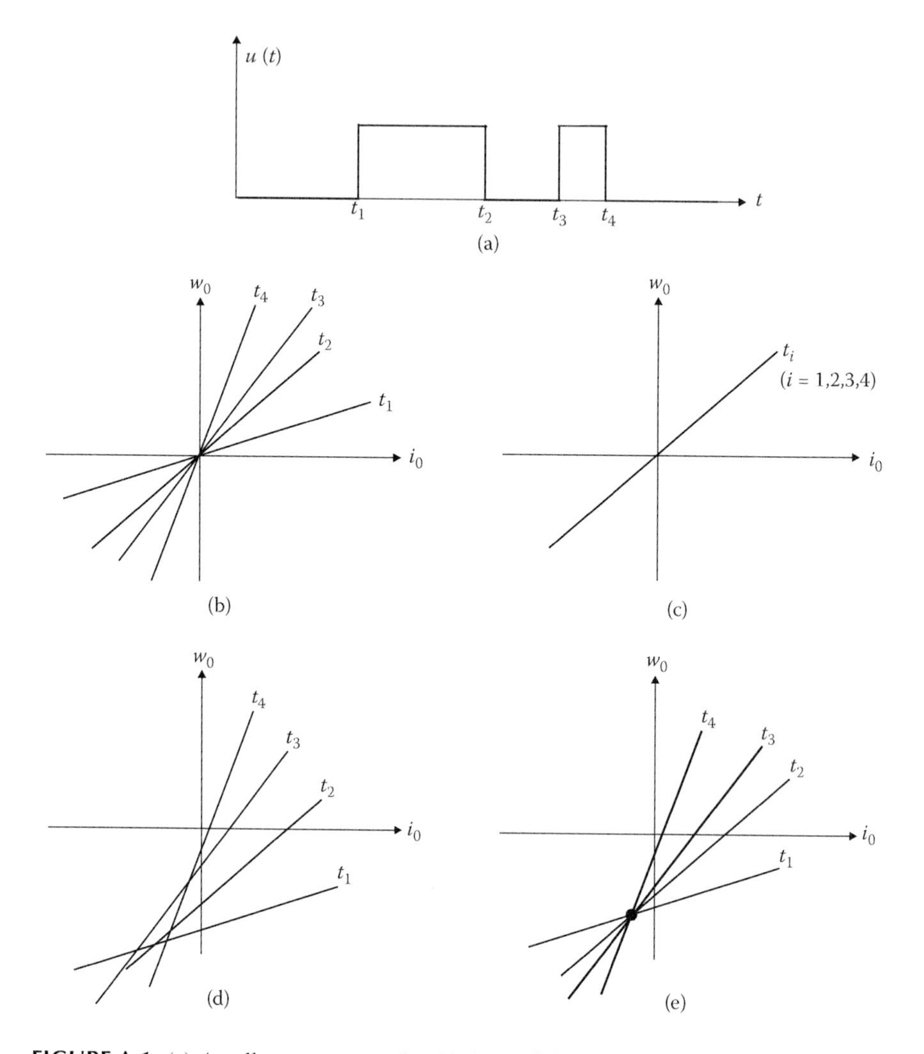

FIGURE A.1 (a) An all-or-none control $u(t)$ that satisfies $u(0) = u(T) = 0$ and switches four times at t_i $(i = 1, 2, 3, 4)$. (b) For the case $\rho_0 = 0$, the i_0 versus w_0 plane under the regular condition (the four lines are different). The line t_i corresponds to Equation (A.32) with t substituted by t_i. (c) For the case $\rho_0 = 0$, the i_0 versus w_0 plane under a singular condition (the four lines overlap). (d) For the case $\rho_0 < 0$, the i_0 versus w_0 plane under the regular condition (the four lines do not have a common intersection). (e) For the case $\rho_0 < 0$, the i_0 versus w_0 plane under a singular condition (the four lines have a common intersection).

the origin is the only intersection, one must have $i_0 = w_0 = 0$. Together with $\rho_0 = 0$, this implies $\rho(t) \equiv 0$ and $u(t) \equiv 0$ is the only locally optimal control (Equation (A.30)). Thus, the present $u(t)$ cannot possibly be optimal, with the condition $\rho_0 = 0$. Theoretically, there may exist a condition in which the four lines are identical (Figure A.1(c)), namely,

$$D(t_1) = D(t_2) = D(t_3) = D(t_4). \tag{A.33}$$

In this case, there may exist a nonzero $\rho(t)$ that makes $u(t)$ "optimal." However, Equation (A.33) itself represents a singular condition. A slight perturbation of any of the parameters in the mathematical model (Equations (2.1, 2.2)), no matter how small, will certainly invalidate Equation (A.33). Therefore, $u(t)$ falls into category (3) defined in Section 3.4 and is thus not a locally optimal control.

- $\rho_0 < 0$. The time point t_i $(i = 1, 2, 3, 4)$ now corresponds to a straight line that intercepts the w_0-axis at $\rho_0 / C(t_i)$ with the slope $D(t_i)$. Under regular conditions the four lines do not have a common intersection (Figure A.1(d)). Thus no (ρ_0, i_0, w_0) value can be chosen to construct a $\rho(t)$ that renders $u(t)$ as an optimal control. Under a singular condition the four lines intersect at the same point (Figure A.1(e)), which may provide a $\rho(t)$ that makes $u(t)$ "optimal." But $u(t)$ again falls into category (3) defined in Section 3.4 and is thus not a locally optimal control.

Therefore, a control that switches four times cannot be optimal. It is obvious that the conclusion can be extended to the cases $N = 6, \cdots, \infty$. Therefore, the only possibility is $N = 2$. That is, an optimal control only switches twice.

For $N = 2$, neither $\rho_0 > 0$ nor $\rho_0 = 0$ leads to optimal solutions, for the reasons given above. Therefore, only $N = 2$ and $\rho_0 < 0$ may lead to optimal solutions. Figure A.2(a) gives an example of optimal $u(t)$; it is clearly a pulse with only one square. Because Pontryagin's maximum principle obtains the necessary conditions for optimality, any optimal $u(t)$ must be a square pulse. But in general a square pulse may not be optimal. If a square pulse satisfies an additional condition (expression (A.35)), then it is also an optimal control. For the normal parameters of the glucose-insulin feedback system, that condition is always satisfied. Therefore, an arbitrary square pulse is a locally optimal control of the glucose-insulin feedback system.

The results in this section are summarized as the following theorem, with the proof given thereafter.

Theorem A.2 *Consider the system described by Equations (A.3–A.5). The control $u(t)$ is locally optimal, if and only if*

(i) $u(t)$ has the form of a square pulse:

$$u(t) = \begin{cases} 0 & \text{for } 0 \le t < t_{\text{on}} \\ u_{\max} & \text{for } t_{\text{on}} \le t < t_{\text{off}} \\ 0 & \text{for } t_{\text{off}} \le t \le T \end{cases} \tag{A.34}$$

where t_{on} and t_{off} can take any values in the range $0 < t < T$; and

(ii) the resultant dynamics satisfy the condition

$$D(t_{\text{off}}) > D(t_{\text{on}}), \tag{A.35}$$

where $D(t)$ is expressed by Eq. (A.28).

Proof. Necessity: $u(t)$ is locally optimal→ conditions (i) and (ii).
If $u(t)$ is a locally optimal control, then it only switches twice in the range $0 < t < T$. Together with the condition $u(0) = u(T) = 0$, one finds that $u(t)$ must have

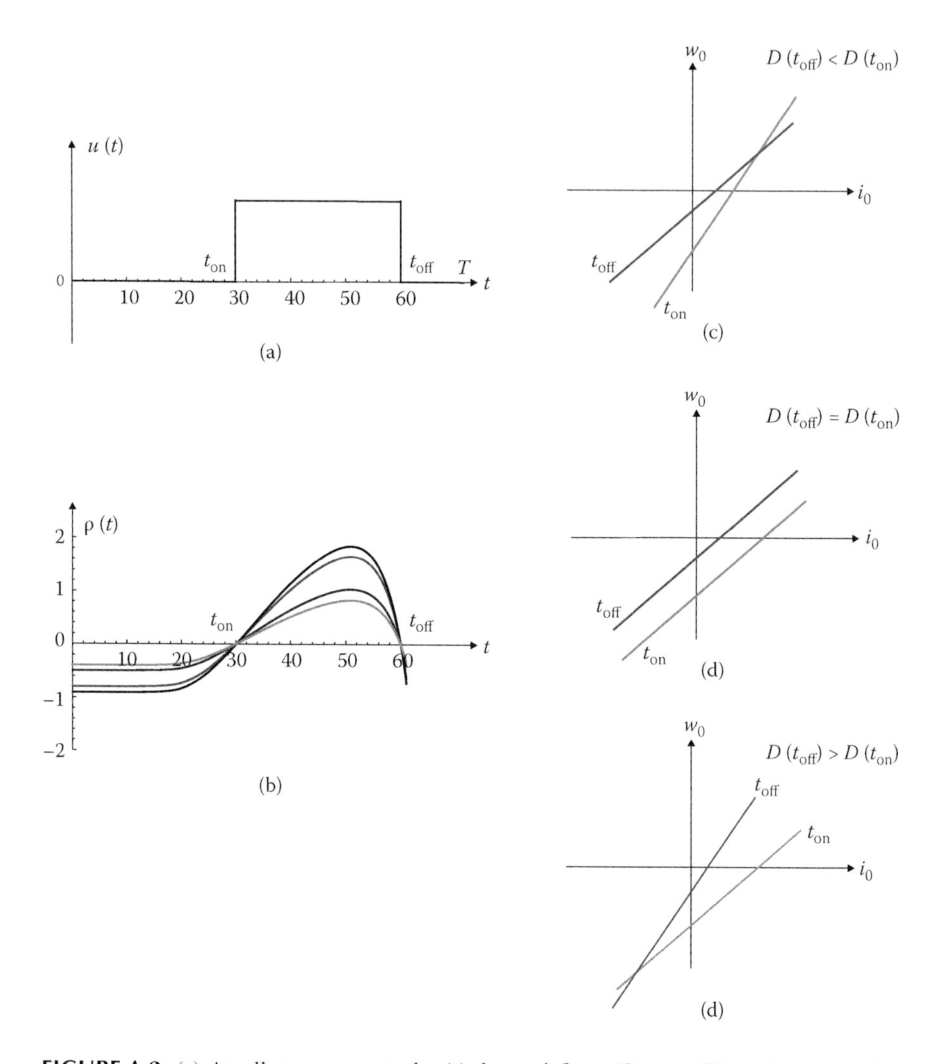

FIGURE A.2 (a) An all-or-none control $u(t)$ that satisfies $u(0) = u(T) = 0$ and switches twice at t_{on} and t_{off}. (b) The switch function $\rho(t)$ that traverses the t-axis at t_{on} and t_{off}. Four examples are shown. (c) The i_0 versus w_0 plane under a regular condition $D(t_{off}) < D(t_{on})$. The two lines intersect at a point with a positive w_0 value, which does not confer an optimal control. (d) The i_0 versus w_0 plane under a singular condition $D(t_{off}) = D(t_{on})$. (e) The i_0 versus w_0 plane under a regular condition $D(t_{off}) > D(t_{on})$. The two lines intersect at a point with a negative w_0 value, which confers an optimal control.

the form described by Equation (A.34) (see Figure A.2(a)). That is, condition (i) must hold.

To prove condition (ii), one can use proof by contradiction. Suppose condition (ii) does not hold (i.e., $D(t_{off}) = D(t_{on})$ or $D(t_{off}) < D(t_{on})$ holds); one needs to prove that $u(t)$ is not a locally optimal control. Note that geometrically $D(t)$ is the slope of the line corresponding to t on the i_0 versus w_0 plane.

- The case $D(t_{\text{off}}) = D(t_{\text{on}})$ is a singular condition, which corresponds to the parallelism of two straight lines on the i_0 versus w_0 plane (Figure A.2(d)). It requires no further consideration.
- The case $D(t_{\text{off}}) < D(t_{\text{on}})$ corresponds to the following scenario: on the i_0 versus w_0 plane, the line corresponding to t_{on} is steeper than the line corresponding to t_{off} (Figure A.2(c)). The two lines intersect at a point (i_0, w_0) with

$$i_0 = \frac{\rho_0}{C(t_{\text{on}})C(t_{\text{off}})}\frac{C(t_{\text{off}}) - C(t_{\text{on}})}{D(t_{\text{off}}) - D(t_{\text{on}})}, \tag{A.36}$$

$$w_0 = \frac{\rho_0}{C(t_{\text{on}})C(t_{\text{off}})}\frac{B(t_{\text{off}}) - B(t_{\text{on}})}{D(t_{\text{off}}) - D(t_{\text{on}})}. \tag{A.37}$$

  Because the function $B(t)$ is an increasing function and $t_{\text{on}} < t_{\text{off}}$, one has $B(t_{\text{off}}) - B(t_{\text{on}}) > 0$. Together with the conditions $\rho_0 < 0$ and $D(t_{\text{off}}) - D(t_{\text{on}}) < 0$, one has $w_0 > 0$. This renders $H(\mathbf{x}(T), \mathbf{X}(T), T, u(T)) = -\eta u_{\max} w_0 < 0$, which implies that $u(t)$ is not a locally optimal control (see Theorem A.1).

Therefore, if $u(t)$ is a locally optimal control, then condition (ii) holds.

Taken together, it has been proved that if $u(t)$ is a locally optimal control, then conditions (i) and (ii) hold.

Sufficiency: conditions (i) and (ii) → $u(t)$ is locally optimal.

The sufficiency is proved if one can find a $\rho(t)$ with $\rho_0 < 0$ that traverses the t-axis precisely at t_{on} and t_{off} and if $H(\mathbf{x}(T), \mathbf{X}(T), T, u(T)) > 0$ holds.

To construct such a $\rho(t)$, one can arbitrarily choose a negative ρ_0 and then determine i_0 and w_0. In fact, the results have been presented by Equations (A.36, A.37). Because the function $B(t)$ is an increasing function, one has $B(t_{\text{off}}) - B(t_{\text{on}}) > 0$. Together with $\rho_0 < 0$ and $D(t_{\text{off}}) - D(t_{\text{on}}) > 0$, one has $w_0 < 0$. Graphically, this corresponds to the intersection of the two lines in the third quadrant (Figure A.2(e)). This renders

$$H(\mathbf{x}(T), \mathbf{X}(T), T, u(T)) = -\eta u_{\max} w_0 > 0,$$

which implies $u(t)$ is a locally optimal control. ■

The control represented by Equation (A.34) is abbreviated 0-$u_{\max}$-0. Because it is completely determined by t_{on} and t_{off}, the control can be denoted $(t_{\text{on}}, t_{\text{off}})$. For example, the 0-$u_{\max}$-0 control in Figure A.2(a) is denoted (30, 60). Because $\rho_0 < 0$ can be arbitrarily chosen, there exist infinitely many $\rho(t)$ that traverse the t-axis at t_{on} and t_{off}. Figure A.2(b) illustrates four such $\rho(t)$, computed by Equation (A.30) with the parameters from Table 3.1.

It should be noted that condition (ii) is usually satisfied in practice. Both $B(t)$ and $C(t)$ are monotonically increasing functions of the time t. However, $B(t)$ usually increases much faster than $C(t)$, due to the term $e^{k\tau}$ and the relatively large value $k = 0.3$ (see Table 3.1). That is, it is almost always that $D(t_{\text{off}}) > D(t_{\text{on}})$. In the search for the globally optimal control, some 8.3×10^5 locally optimal controls were enumerated, and condition (ii) holds for all these controls.

A.5 SEARCH FOR THE GLOBAL OPTIMA

Usually, there is one and only one global optimum because the word "global" refers to "the best." For the present problem, the controls are partitioned according to the utilization rate η. Therefore, there are infinitely many globally optimal controls, each corresponding to a specific utilization rate η. That is, the optimization is restricted to all the controls having the same η value; each η has its own globally optimal control.

A global optimum must be a local one in the first place. Because the locally optimal controls can be characterized by only two parameters, t_{on} and t_{off}, one can enumerate all the relevant (t_{on}, t_{off}) pairs to select the best ones. Figure 3.5(a) illustrates the area of all the meaningful (t_{on}, t_{off}) pairs. There are 8.3×10^5 points in the area, up to a step size 0.05 min for both the t_{on} dimension and the t_{off} dimension. The area is constructed as follows.

- The left boundary (the line da) signifies that the scan of t_{on} begins at 1 minute.
- The right boundary (the line bc) corresponds to $t_{on} = 70.65$, the time when the uncontrolled insulin level (i.e., the curve $\hat{I}(t)$ in Figure 3.5(b)) reaches the maximum. To see why $t_{on} > 70.65$ is impractical, let $t_{on} = 70.65 + \Delta R$, where ΔR is positive. As shown in Figure 3.5(b), this t_{on} corresponds to the insulin concentration $\hat{I}(70.65 + \Delta R)$, which equals $\hat{I}(70.65 - \Delta L)$. Because the control is actually mediated by insulin, the control should have turned on at time $70.65 - \Delta L$, instead of $70.65 + \Delta R$. That is, $t_{on} > 70.65$ is meaningless.
- The lower boundary (the line ab) is the diagonal, which corresponds to the fact $t_{off} > t_{on}$.
- The upper boundary (the line cd) corresponds to the limit condition $t_{off} = T$. That is, the control must have switched off before homeostasis is reached; otherwise $u(T) = u_{max}$ would result, which would violate the requirement $u(T) = 0$. Figure 3.5(c) illustrates control with $t_{off} = T$, as an example.

For each point, the homeostasis time T and the utilization rate η are calculated in the following way. One first constructs a 0-u_{max}-0 control $u(t)$ that switches on at time t_{on} and switches off at time t_{off}. The control $u(t)$ is then used to run the mathematical model to obtain $G(t)$ and $I(t)$. Because the states will never reach G_0 and I_0 exactly, the values $(1+\varepsilon)G_0$ and $(1+\varepsilon)I_0$ are set as the indicators of homeostasis, where $\varepsilon = 0.1$. In other words, the homeostasis time T is defined as the first time for which $G(t) \leq 1.1G_0$ and $I(t) \leq 1.1I_0$ hold forever. To obtain the utilization rate, one just uses the formula $\eta = (t_{off} - t_{on})/T$, the duty cycle of a 0-$u_{max}$-0 control. In this way, a dataset is formed containing the T and η values of each point in the area.

To obtain the global optimal control for a particular η, one needs only to extract from the dataset those controls whose utilization rates are within a narrow range $\eta \pm \Delta\eta$, where the tolerance $\Delta\eta$ accounts for the fact that few, if any, controls have a utilization rate exactly η. Among the extracted controls, the one with the smallest T is the globally optimal control for η.

B Derivation of the Master Equation of AKT Activation

In this appendix, we start from the steady-state equations of the PI3K-AKT-TOR pathway:

$$-a_1[\mathrm{AKT}][\mathrm{E}_1] + d_1[\mathrm{AKT{:}\ E}_1] + k_2[\mathrm{pAKT : E}_2] = 0, \tag{B.1}$$
$$a_1[\mathrm{AKT}][\mathrm{E}_1] - (d_1 + k_1)[\mathrm{AKT{:}\ E}_1] = 0, \tag{B.2}$$
$$-a_2[\mathrm{pAKT}][\mathrm{E}_2] + d_2[\mathrm{pAKT{:}\ E}_2] + k_1[\mathrm{AKT : E}_1] = 0, \tag{B.3}$$
$$a_2[\mathrm{pAKT}][\mathrm{E}_2] - (d_2 + k_2)[\mathrm{pAKT{:}\ E}_2] = 0, \tag{B.4}$$
$$I + (\Phi - \Psi[\mathrm{pIRS}])[\mathrm{pAKT}] - \delta[\mathrm{pIRS}] = 0, \tag{B.5}$$
$$[\mathrm{AKT}] + [\mathrm{pAKT}] = A_{\mathrm{max}}, \tag{B.6}$$
$$[\mathrm{E}_1] + [\mathrm{AKT{:}\ E}_1] = E_{1T}, \tag{B.7}$$
$$[\mathrm{E}_2] + [\mathrm{pAKT{:}\ E}_2] = E_{2T}, \tag{B.8}$$
$$\beta[\mathrm{pIRS}] = E_{1T}, \tag{B.9}$$

to derive the master equation of AKT activation:

$$G(I, A, \mathbf{p}) = 0, \tag{B.10}$$

where $G(I, A, \mathbf{p})$ abbreviates a long expression:

$$\begin{aligned} G = \theta A^3 &+ \Big((K_1 + K_2)\phi - (K_1 + 1)\theta + \beta\gamma I - 1 \Big) A^2 \\ &+ \Big(K_1 + 1 + (K_2 - 1)\beta\gamma I - K_2\phi \Big) A - K_2\beta\gamma I \end{aligned} \tag{B.11}$$

and where

$$\begin{aligned} \mathbf{p} &= (\phi, \theta, \beta, \gamma, K_1, K_2), \\ \gamma &= \frac{k_1}{\delta k_2 E_{2T}}, \\ \phi &= \Phi\beta\gamma A_{\mathrm{max}}, \\ \theta &= (\Phi\beta\gamma - \Psi/\delta) A_{\mathrm{max}}, \\ K_1 &= \frac{K_{m1}}{A_{\mathrm{max}}} = \frac{d_1 + k_1}{a_1 A_{\mathrm{max}}}, \\ K_2 &= \frac{K_{m2}}{A_{\mathrm{max}}} = \frac{d_2 + k_2}{a_2 A_{\mathrm{max}}}. \end{aligned}$$

From Equations (B.2, B.7), one obtains

$$[\text{AKT: E}_1] = \frac{E_{1T}[\text{AKT}]}{K_{m1} + [\text{AKT}]}. \tag{B.12}$$

Because of Equation (B.6), one further obtains

$$[\text{AKT: E}_1] = \frac{E_{1T}(X_T - [\text{pAKT}])}{K_{m1} + X_T - [\text{pAKT}]}. \tag{B.13}$$

From Equations (B.4, B.8), one obtains

$$[\text{pAKT: E}_2] = \frac{E_{2T}[\text{pAKT}]}{K_{m2} + [\text{pAKT}]}. \tag{B.14}$$

The addition of Equation (B.1) and Equation (B.2) yields

$$k_1[\text{AKT: E}_1] = k_2[\text{pAKT: E}_2]. \tag{B.15}$$

Combining Equations (B.13, B.14, B.15), one obtains

$$\frac{k_1 E_{1T}(X_T - [\text{pAKT}])}{K_{m1} + X_T - [\text{pAKT}]} = \frac{k_2 E_{2T}[\text{pAKT}]}{K_{m2} + [\text{pAKT}]}. \tag{B.16}$$

From Equations (B.5, B.9), one obtains

$$I + (\Phi - \Psi E_{1T}/\beta)[\text{pAKT}] = \delta E_{1T}/\beta. \tag{B.17}$$

Combining Equations (B.16, B.17) by eliminating E_{1T}, one obtains a cubic equation (third degree of [pAKT]). By using $A = [\text{pAKT}]/X_T$ to denote the percentage of the activated AKT, one obtains Equation (B.10), the master equation of AKT activation, namely, Equation (9.23) in the main text.

C Singularity Theory

In Chapter 9, we hypothesized that different *kinds* of response curve may correspond to different physiologic/pathologic phenotypes of the PI3K-AKT-TOR pathway. The three quantities I, A, and $\mathbf{p}$ are constrained by the master equation of AKT activation:

$$G(I, A, \mathbf{p}) = 0.$$

According to the master equation, every point $\mathbf{p}$ in the parameter space is associated with a shape of the response curve $A(I)$. The parameter space can thus be divided into different regions, with each region including all the points whose response curves are of the same kind, which is qualitatively different from the kind of response curves of another region. That is, each region corresponds to a unique phenotype, and such a demarcation of the parameter space would advance our understanding of the PI3K-AKT-TOR pathway. Theoretically speaking, the demarcation can be accomplished after all the points in the parameter space are investigated and classified. However, this approach is very time consuming; it is even impossible if the parameter space is too large.

Singularity theory aims to avoid unnecessary computations. The key idea is that there must be boundaries between different regions in the parameter space. If the boundaries are first determined, then the demarcation has already been accomplished. In other words, it is actually unnecessary to investigate every point and then classify the points into different regions. One may use an analogy with drawing an object. By tracing out the boundary between the object and the air, the object is already drawn. One actually need not depict every point of the object.

C.1 NONPERSISTENT POINTS

The boundary points in the parameter space are mathematically special, because they define the qualitative transition from one kind of response to another. Such points are called *nonpersistent points*. To explain this concept, we use a two-dimensional parameter space $\mathbf{p} = [\alpha, \beta]$ (Figure C.1). Consider a boundary (e.g., the curve $\beta = \alpha^3$ as shown in Figure C.1) that divides the space into two regions A and B. Points in the two regions are drawn in different colors (blue versus green), because their corresponding response curves are qualitatively different. The boundary is colored in black, because points on the boundary correspond to yet another kind of response curve. Consider a point $\mathbf{p}$ in the region B. An infinitesimal perturbation of $\mathbf{p}$ leads to a new point $\mathbf{p} + \delta\mathbf{p}$. No matter what $\delta\mathbf{p}$ is, point $\mathbf{p} + \delta\mathbf{p}$ always stays in region B. The perturbation does not cause any qualitative change. We can thus say that $\mathbf{p}$ is a *persistent* point. Now let's consider a point $\mathbf{p}_0$ on the boundary. It is a nonpersistent point, because an infinitesimal $\delta\mathbf{p}$ would perturb the point away from the boundary ($\mathbf{p}_0 + \delta\mathbf{p}$ is either in the region A or in the region B).

Singularity theory has revealed four kinds of nonpersistent points: simple bifurcation point, isola center, hysteresis point, and double limit point. For a simpler

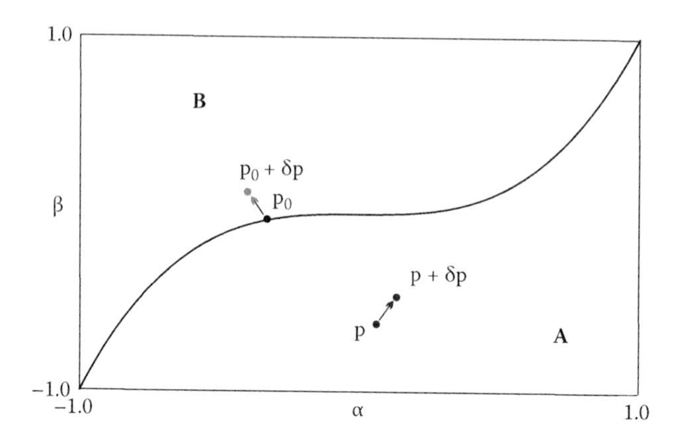

FIGURE C.1 The two-dimensional space $\mathbf{p} = [\alpha, \beta]$ that is divided into two regions A and B. See color insert.

elucidation of these nonpersistent points, we assume the vector of parameters $\mathbf{p} = [\alpha]$ contains only one parameter α. Therefore, the parameter space is just the α-axis, and each point on the α-axis corresponds to a response curve $A(I)$ (a general curve here, not necessarily involving the PI3K-AKT-TOR pathway). Figure C.5 gives an example of the qualitative change of $A(I)$ as the value of α changes. The three quantities I, A, and α are constrained by the equation

$$G(I, A, \alpha) = 0, \tag{C.1}$$

or $G = 0$ for short. It is a general equation, not necessarily representing the master equation of AKT activation.

Equation (C.1) is just an equation for steady states. Additional equations are needed to determine the nonpersistent points. These equations are collectively called the normal form equations. Different nonpersistent points are characterized by different kinds of normal form equations.

C.1.1 Simple Bifurcation Points

Simple bifurcation points satisfy the following normal form equations:

$$\begin{aligned} G &= 0, \\ \frac{\partial G}{\partial A} &= 0, \\ \frac{\partial G}{\partial I} &= 0. \end{aligned} \tag{C.2}$$

These three equations can determine the three unknowns I, A, and α. Let I_0, A_0, α_0 be the solution. In particular, $\alpha = \alpha_0$ is the so-called nonpersistent point. It is the boundary point that divides the α-axis into two parts: $\alpha < \alpha_0$ and $\alpha > \alpha_0$ (see Figure C.2). By substituting $\alpha = \alpha_0$ back into $G(I, A, \alpha) = 0$, one obtains the specific response curve $A(I)$ that corresponds to α_0 (Figure C.2(b)). The response

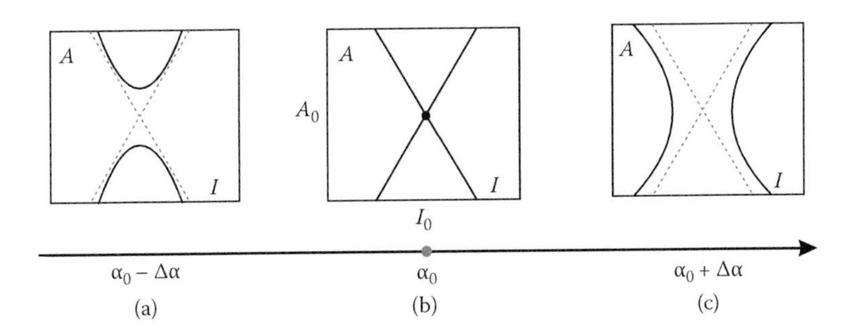

FIGURE C.2 Simple bifurcation point α_0. The change of the parameter α around α_0 causes qualitative changes of the corresponding curves.

curve consists of two straight lines crossing at the point (I_0, A_0). A perturbation of α_0 turns the two straight lines into two branches of a hyperbola, which separates either in the vertical direction (Figure C.2(a)) or in the horizontal direction (Figure C.2(c)).

It should be noted that Equation (C.2) is not exclusive for simple bifurcation points. Isola centers also satisfy Equation (C.2). For this reason, both simple bifurcation points and isola centers are denoted by $\mathscr{B}$ in this book. The differences between simple bifurcation points and isola centers consist in higher-order terms $\partial^2 G/\partial A^2$, $\partial^2 G/\partial I^2$, $\partial^2 G/(\partial G \partial I)$, ..., which are not reflected by Equation (C.2).

C.1.2 Isola Centers

Isola centers also satisfy the three equations in (C.2). Let I_0, A_0, α_0 be the solution to the three equations. In particular, $\alpha = \alpha_0$ is the nonpersistent point. It is the boundary point that divides the α-axis into two parts: $\alpha < \alpha_0$ and $\alpha > \alpha_0$ (see Figure C.3). By substituting $\alpha = \alpha_0$ back into $G(I, A, \alpha) = 0$, one obtains the specific response curve $A(I)$ that corresponds to α_0 (Figure C.3(b)), which is quite different from Figure C.2(b), the response curve of a simple bifurcation point. For the case of a simple bifurcation point, the function G is such that the equation $G(I, A, \alpha_0) = 0$ results in two crossing lines as $A(I)$, as already described. For the case of isola centers, the function G is so special that the equation $G(I, A, \alpha_0) = 0$ has only one solution: $I = I_0$ and $A = A_0$. As a consequence, Figure C.3(b) contains only one point (I_0, A_0) as "the response curve" associated with α_0. A perturbation of α_0 either leads to no solution (Figure C.3(a)) or leads to a circle (Figure C.3(c)).

Isola centers are also denoted by $\mathscr{B}$.

C.1.3 Double Limit Points

Simple bifurcation points, isola centers, and hysteresis points (see below) all involve only one set of (I_0, A_0, α_0). This is not true for the case of double limit points, which have two solutions, (I_0, A_1, α_0) and (I_0, A_2, α_0), for the following normal form equations:

$$G = 0$$

$$\frac{\partial G}{\partial A} = 0 \tag{C.3}$$

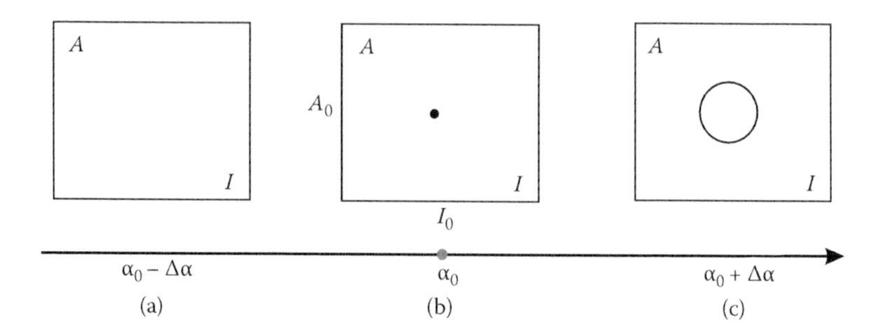

FIGURE C.3 Isola center α_0. The change of the parameter α around α_0 causes qualitative changes of the corresponding curves.

The point $\alpha = \alpha_0$ is the nonpersistent point. It is the boundary point that divides the α-axis into two parts: $\alpha < \alpha_0$ and $\alpha > \alpha_0$ (see Figure C.4). By substituting $\alpha = \alpha_0$ back into $G(I, A, \alpha) = 0$, one obtains the specific response curve $A(I)$ that corresponds to α_0 (Figure C.4(b)). This response curve is a hyperbola with two branches. The vertices of the two branches have the same I value (I_0) but different A values (A_1 and A_2). A perturbation of α_0 either pushes the two branches apart (Figure C.4(a)) or pulls the two branches closer (Figure C.4(c)).

Double limit points are denoted by $\mathscr{D}$.

C.1.4 Hysteresis Points

Hysteresis points satisfy the following normal form equations:

$$\begin{aligned} G &= 0 \\ \frac{\partial G}{\partial A} &= 0 \\ \frac{\partial^2 G}{\partial A^2} &= 0 \end{aligned} \tag{C.4}$$

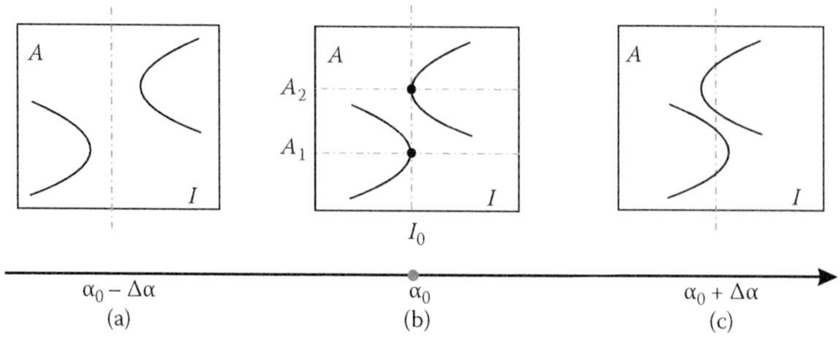

FIGURE C.4 Double limit point α_0. The change of the parameter α around α_0 causes qualitative changes of the corresponding curves.

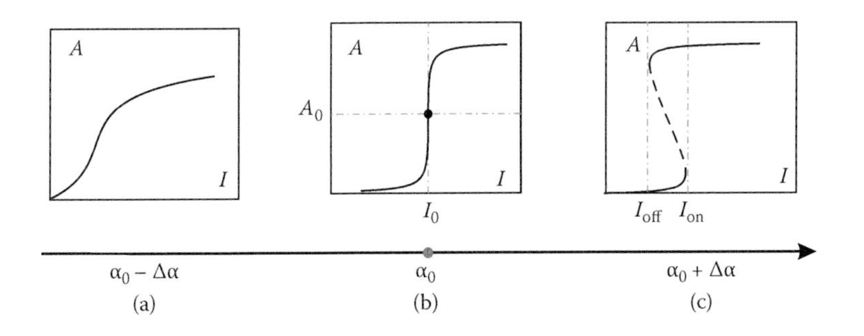

FIGURE C.5 Hysteresis point α_0. The change of the parameter α around α_0 causes qualitative changes of the corresponding curves.

These three equations can determine the three unknowns I, A, and α. Let I_0, A_0, α_0 be the solution. In particular, $\alpha = \alpha_0$ is the nonpersistent point. It is the boundary point that divides the α-axis into two parts: $\alpha < \alpha_0$ and $\alpha > \alpha_0$ (see Figure C.5). By substituting $\alpha = \alpha_0$ back into $G(I, A, \alpha) = 0$, one obtains the specific response curve $A(I)$ that corresponds to α_0 (Figure C.5(b)). This response curve is characterized by the vertical tangent at the point (I_0, A_0). It is thus a discontinuous function.

A perturbation of α_0 leads to either a monotonically increasing function (Figure C.5(a)) or an S–shaped curve (Figure C.5(c)). This S–shaped curve is not a function. To be a function, one preimage must correspond to at most one image. But in the range $[I_{off}, I_{on}]$, one preimage corresponds to three images. The middle image is generally unstable and is thus represented by a dashed line. The other two images are stable. The S–shaped curve is thus called a toggle switch. The qualitative transition around α_0 is clear: as α changes from $\alpha_0 - \Delta\alpha$ to α_0 and then to $\alpha_0 + \Delta\alpha$, the curve changes from a continuous function to a discontinuous function and then to a nonfunction. The toggle switch has two switch points, I_{off} and I_{on}, indicating that the route of switch-on is different from that of switch-off (hysteresis). This property might be the etymology of hysteresis points.

Hysteresis points are denoted by $\mathcal{H}$.

C.2 IRREVERSIBILITY POINTS

Unlike nonpersistent points, which are quite general concepts, irreversibility points are defined solely for the present biological problem (i.e., I represents the level of growth factor and A represents the percentage of [pAKT]). To explain irreversibility points, the response curves in Figure C.5 are reproduced in Figure C.6, which includes two additional curves (Figure C.6(d, e)).

As α increases, the response curve gradually deforms into a bistable switch that is characterized by two thresholds, I_{on} and I_{off}. The curve in Figure C.6(c) can be called a toggle switch because A toggles between about A_{max} (when I exceeds I_{on}) and about 0 (when I drops below I_{off}). As α further increases, the deformation of the curve continues, and the values of I_{on} and I_{off} continue to change. Finally, I_{off} becomes negative, and the response curve can no longer be called a toggle switch

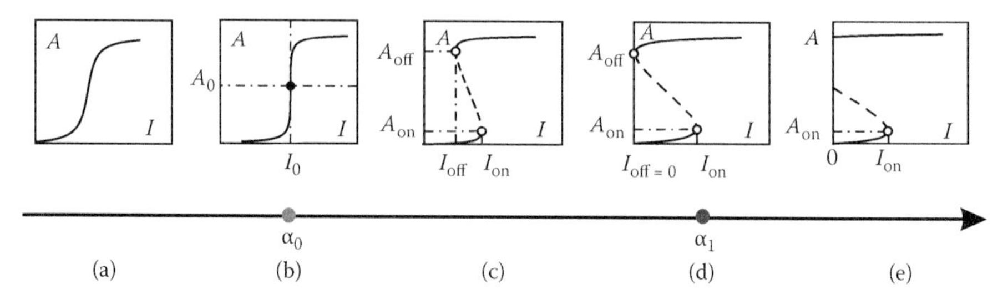

FIGURE C.6 Irreversibility point α_1. The increase of the parameter α turns the toggle switch into an irreversible switch. The change occurs at $\alpha = \alpha_1$, which corresponds to the condition $I_{off} = 0$.

(Figure C.6(e)). It is called an irreversible switch in the following sense. It is easy for A to switch to about A_{max} (simply by increasing I over I_{on}). However, the activation cannot be reversed by decreasing I: once A reaches about A_{max}, it maintains at about A_{max}. Even if I reduces to 0, A is still about A_{max}. To bring A back to about 0, I has to be smaller than the negative I_{off}, which is impossible because I, as the level of growth factor, must be positive.

The condition $I_{off} = 0$ (Figure C.6(d)) marks the boundary between the toggle switch and the irreversible switch. The corresponding α value, α_1, is thus called the irreversibility point. Note that the irreversibility point is not a nonpersistent point, because the toggle switch and the irreversibility switch are qualitatively the same. They are both bistable switches.

Irreversibility points satisfy the following normal form equations:

$$\begin{aligned} G &= 0, \\ \frac{\partial G}{\partial A} &= 0, \\ I &= 0, \end{aligned} \tag{C.5}$$

where the first two equations characterize the turning points of the responsive curve: (I_{on}, A_{on}) and (I_{off}, A_{off}). The last equation imposes that the point (I_{off}, A_{off}) must be located on the vertical axis: $I_{off} = 0$.

The following facts are worth noting. Simple bifurcation points, isola centers, and hysteresis points involve only one set of (I_0, A_0, α_0). Double limit points involve two sets, (I_0, A_1, α_0) and (I_0, A_2, α_0), but the difference is only at the A-value. Irreversibility points involve two sets, $(I_{on}, A_{on}, \alpha_1)$ and $(I_{off}, A_{off}, \alpha_1)$, and the differences are at both I- and A-values.

D Analysis of the Simplified Master Equation

In this appendix, singularity theory is applied to the simplified master equation of form II:

$$\tilde{G}\left(\tilde{I}, A, \tilde{\mathbf{p}}\right) = 0, \tag{D.1}$$

where

$$\tilde{G} = \tilde{\theta}A^3 + \left((K_2 - 1)\tilde{\theta} + \tilde{I} - 1\right)A^2$$
$$+\left(K_1 + 1 + (K_2 - 1)\tilde{I} - K_2\tilde{\theta}\right)A - K_2\tilde{I},$$

and

$$\tilde{\mathbf{p}} = (\tilde{\theta}, K_1, K_2),$$
$$\tilde{\theta} = \alpha\beta,$$
$$\tilde{I} = \beta\gamma I.$$

Based on Equation (D.1), we set out to identify all the nonpersistent points and irreversibility points in the space of $\tilde{\mathbf{p}}$.

D.1 NONPERSISTENT POINTS

D.1.1 Simple Bifurcation Points and Isola Centers

Simple bifurcation points and isola centers are special points in the space of $\tilde{\mathbf{p}}$ that satisfy

$$\tilde{G} = \frac{\partial \tilde{G}}{\partial A} = \frac{\partial \tilde{G}}{\partial \tilde{I}} = 0.$$

Because $\partial \tilde{G}/\partial \tilde{I} = A^2 + (K_2 - 1)A - K_2$, the condition $\partial \tilde{G}/\partial \tilde{I} = 0$ implies

$$A^2 + (K_2 - 1)A - K_2 = 0,$$

which has two solutions: $A = 1$ and $A = -K_2$. The solution $A = -K_2$ is obviously meaningless ([pAKT] cannot be negative). The solution $A = 1$ is also absurd, because it renders $K_1 = 0$ after substituting $A = 1$ into the equation $\tilde{G} = 0$. Therefore, there are no simple bifurcation points and no isola centers in the space of $\tilde{\mathbf{p}}$. That is, $\mathscr{B} = \phi$.

D.1.2 Double Limit Points

According to lemma 7.1 of [12], polynomial equations of degree 3 or less do not have double limit points. Because Equation (D.1) is a third degree equation, there are no double limit points in the space of $\tilde{\mathbf{p}}$. That is, $\mathscr{D} = \phi$.

D.1.3 Hysteresis Points

Hysteresis points are special points in the space of $\tilde{\mathbf{p}}$ that satisfy

$$\tilde{G} = \frac{\partial \tilde{G}}{\partial A} = \frac{\partial^2 \tilde{G}}{\partial A^2} = 0.$$

After eliminating $\tilde{I}$ and A, the three equations above reduce to one equation:

$$\tilde{a}(K_1, K_2)\tilde{\theta}^3 + \tilde{b}(K_1, K_2)\tilde{\theta}^2 + \tilde{c}(K_1, K_2)\tilde{\theta} + \tilde{d}(K_1, K_2) = 0 \tag{D.2}$$

where

$$\tilde{a} = (1 + K_2)^6, \tag{D.3}$$

$$\tilde{b} = -3(1 + K_2)^4(K_1 + K_2), \tag{D.4}$$

$$\begin{aligned}\tilde{c} = {} & 3K_2^2(1 + K_2)^2 + 3K_1^2\left(1 - 7K_2 + K_2^2\right) \\ & +3K_1K_2(2K_2^2 - 5K_2 - 7),\end{aligned} \tag{D.5}$$

$$\tilde{d} = -(K_1 + K_2)^3. \tag{D.6}$$

Equation (D.2) constrains the relationship among the three parameters $\tilde{\theta}$, K_1, and K_2. It is a third degree equation of $\tilde{\theta}$, with all the coefficients being functions of K_1 and K_2. According to the *fundamental theorem of algebra*, a degree n polynomial equation has exactly n roots in the complex domain. Equation (D.2) thus has three roots of $\tilde{\theta}$. However, it is possible that some roots have nonvanishing imaginary parts; some roots are negative. These roots are all meaningless. A meaningful root must be a positive, real number. A natural question is: does Equation (D.2) have meaningful roots for every (K_1, K_2) pair? If yes, how many of them? It is possible that some areas of (K_1, K_2) correspond to three meaningful roots, some areas correspond to two meaningful roots, and some areas correspond to just one meaningful root.

To answer this question, it seems that one has to repeat the following tedious cycle: sampling the values of K_1 and K_2, calculating the corresponding values of $\tilde{\theta}$, and testing the meaningfulness of the obtained $\tilde{\theta}$ values. Here mathematics again shows its power—enumeration and numerical computations are actually not required. One can prove that there is one and only one meaningful solution, no matter what K_1 and K_2 are. To show this, one can symbolically compute the discriminant of Equation (D.2),

$$\Delta = \tilde{b}^2\tilde{c}^2 - 4\tilde{a}\tilde{c}^3 - 4\tilde{b}^3\tilde{d} - 27\tilde{a}^2\tilde{d}^2 + 18\tilde{a}\tilde{b}\tilde{c}\tilde{d}, \tag{D.7}$$

which has a long expression. After factorization, the discriminant is found to be negative:

$$\begin{aligned}\Delta = {} & -19{,}683K_1^2K_2^2(1 + K_2)^6\Big(K_1^2(K_2 - 1) \\ & +K_2(1 + K_2)^2 + K_1\left(2K_2^2 + K_2 - 1\right)\Big)^2 \\ < {} & 0.\end{aligned} \tag{D.8}$$

Therefore, Equation (D.2) has two nonreal complex conjugate roots (meaningless) and one real root. If the real root is negative, then there is no meaningful root and thus no hysteresis point. To judge the positivity of the real root, one can examine the sign of $\tilde{a}\tilde{d}$. Because

$$\tilde{a}\tilde{d} = -(K_1 + K_2)^3(1 + K_2)^6 < 0,$$

the real root must be positive. Therefore, Equation (D.2) always has one and only one positive root of $\tilde{\theta}$, no matter what K_1 and K_2 are (even if they are negative). That is, a given set of K_1 and K_2 values corresponds to one and only one positive $\tilde{\theta}$ value. As K_1 and K_2 change, the positive value of $\tilde{\theta}$ changes accordingly, tracing out a surface in the space of $\tilde{\mathbf{p}}$.

Let $\tilde{\theta} = \tilde{\Theta}_0(K_1, K_2)$ denote the surface. The analytical expression of $\tilde{\Theta}_0(K_1, K_2)$ is

$$\tilde{\Theta}_0 = -\frac{1}{3}\left(\frac{\tilde{b}}{\tilde{a}} + \sqrt[3]{\frac{S+\sqrt{T}}{2}} + \sqrt[3]{\frac{S-\sqrt{T}}{2}}\right), \tag{D.9}$$

where $\tilde{a}$ and $\tilde{b}$ have been expressed in Equations (D.3) and (D.4),

$$S = 2\tilde{b}^3/\tilde{a}^3 - 9\tilde{b}\tilde{c}/\tilde{a}^2 + 27\tilde{d}/\tilde{a},$$
$$T = S^2 - 4\left(\tilde{b}^2/\tilde{a}^2 - 3\tilde{c}/\tilde{a}\right)^3.$$

This surface is drawn with the green color in Figure 9.3. It contains all the hysteresis points.

D.1.4 Qualitative Transition

We have shown that for the simplified master equation, hysteresis points are the only nonpersistent points, and these points constitute a single surface that divides the parameter space in half (Figure D.1). Therefore, the qualitative transition of the response curve is a scenario as simple as the one shown in Figure C.5. For a point below the surface in Figure D.1, the corresponding response curve is a monotonically increasing function, as the one shown in Figure C.5(a). For a point on the surface, the response curve is a function with a vertical tangent, as the one shown in Figure C.5(b). For a point above the surface, the response curve is a bistable switch, as the one shown in Figure C.5(c).

D.2 IRREVERSIBILITY POINTS

Irreversibility points are special points in the space of $\tilde{\mathbf{p}}$ that satisfy

$$\tilde{G} = \frac{\partial \tilde{G}}{\partial A} = \tilde{I} = 0 \tag{D.10}$$

and that correspond to the coordinates $(\tilde{I}_{\text{on}}, A_{\text{on}})$ and $(\tilde{I}_{\text{off}}, A_{\text{off}})$. Because Equation (D.10) itself does not determine which of its solutions correspond to "on" and which to "off," we first use $(\tilde{I}_{\text{m}}, A_{\text{m}})$ to represent both of them, with m = "on" or "off." Equation (D.10) has three solutions.

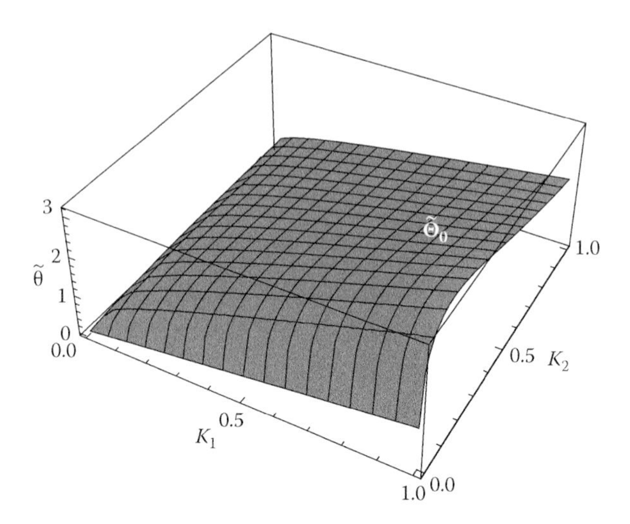

FIGURE D.1 The surface $\tilde{\Theta}_0$ (K_1, K_2) consists of all the hysteresis points. It divides the space into two regions. The region above (below) the surface corresponds to the bistable switch (the monotone type).

D.2.1 Solution 3

This solution constitutes a surface in the space of $\tilde{\mathbf{p}}$:

$$\tilde{\theta} = \tilde{\Theta}_3(K_1, K_2),$$

where

$$\tilde{\Theta}_3 = \frac{1 + K_1}{K_2}. \tag{D.11}$$

The corresponding turning point is $(I_m, A_m) = (0, 0)$. The corresponding response curve is shown in Figure D.2, from which one sees that m actually stands for "on," but not "off." Therefore, this solution does not represent the irreversibility point, which should correspond to $\tilde{I}_{off} = 0$.

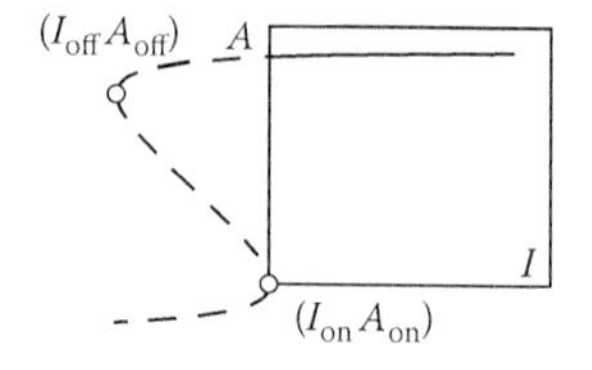

FIGURE D.2 The response curve corresponding to solution 1 of Equation (D.10), which is characterized by $I_{on} = 0$ and $A_{on} = 0$.

D.2.2 Solution 2

This solution constitutes a surface in the space of $\tilde{\mathbf{p}}$:

$$\tilde{\theta} = \tilde{\Theta}_2(K_1, K_2),$$

where

$$\tilde{\Theta}_2 = \frac{1 + K_2 + 2K_1 - 2\sqrt{K_1^2 + K_1 K_2 + K_1}}{(1 + K_2)^2}.$$

The corresponding turning point is (I_m, A_m) with $I_m = 0$ and

$$A_m = 1 + K_1 + \sqrt{K_1^2 + K_1 K_2 + K_1}.$$

This set of solutions is apparently meaningless, because $A_m > 1$ ([pAKT] is more than 100% of the total AKT).

D.2.3 Solution 1

This solution constitutes a surface in the space of $\tilde{\mathbf{p}}$:

$$\tilde{\theta} = \tilde{\Theta}_1(K_1, K_2),$$

where

$$\tilde{\Theta}_1 = \frac{1 + K_2 + 2K_1 + 2\sqrt{K_1^2 + K_1 K_2 + K_1}}{(1 + K_2)^2}. \tag{D.12}$$

The corresponding turning point is (I_m, A_m) with $I_m = 0$ and

$$A_m = 1 + K_1 - \sqrt{K_1^2 + K_1 K_2 + K_1}.$$

This solution is meaningful, because $A_m < 1$. The surface $\tilde{\theta} = \tilde{\Theta}_1(K_1, K_2)$ is drawn in Figure D.3, which contains all the irreversibility points. The points above the

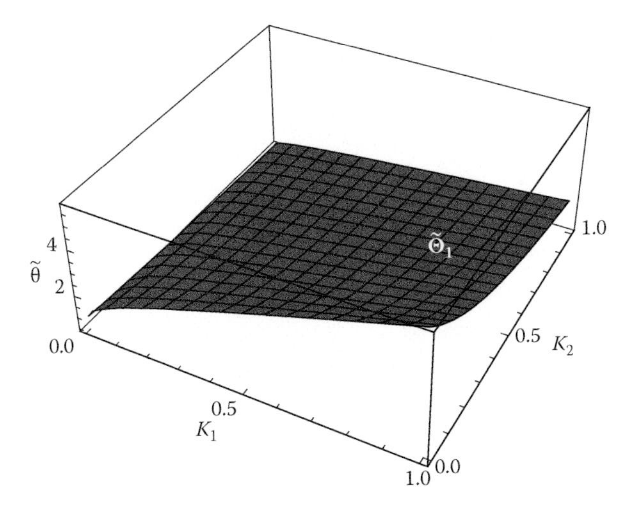

FIGURE D.3 The surface $\tilde{\Theta}_1(K_1, K_2)$ consists of all the irreversible points. It divides the space into two regions. The region above the surface corresponds to irreversible switches.

surface correspond to irreversible switches. The points below the surface correspond to toggle switches or monotonically increasing functions. In Figure 9.3, the red surface $\tilde{\theta} = \tilde{\Theta}_1(K_1, K_2)$ and the green surface $\tilde{\theta} = \tilde{\Theta}_0(K_1, K_2)$ are placed together. One sees that the red surface is above the green surface.

E Analysis of the Master Equation

Thus far, most of our conclusions have been based on the simplified master equation (Equation (9.37)). It is thus important to investigate whether or not the simplified master equation approximates well the full master equation (Equation (9.23)). In this appendix, we aim to apply singularity analysis to the full master equation, and to see if there exist new dynamical behaviors that are unique to the full model. To simplify our analysis, we let $K_1 = K_2 = K$. We also fix $\beta\gamma = 1$ so that $\beta\gamma I$ becomes I (alternatively, we can use $\tilde{I}$ to replace $\beta\gamma I$).

The primary difference between the master equation and the simplified master equation is as follows. In the simplified equation, it is assumed that the negative feedback directly cancels out the positive feedback, resulting in a single parameter $\tilde{\theta}$ that represents the net strength of positive and negative feedback. In the full master equation, the two are not strictly opposing to each other, which necessitates two parameters θ and ϕ to characterize them, where ϕ is the strength of the positive feedback and θ is a quantity similar to the net strength of both. Therefore, by fixing the parameter ϕ, we can analyze the full master equation, similar to what we did for the simplified one. This leads to the following expression:

$$G(I, A, \theta, K|\phi) = 0, \tag{E.1}$$

where

$$\begin{aligned} G = {} & \theta A^3 + (2K\phi - (K+1)\theta + I - 1)A^2 \\ & +(K + 1 + (K-1)I - K\phi)A - KI. \end{aligned}$$

Based on Equation (E.1), we set out to identify all the nonpersistent points and irreversibility points in the space of $\mathbf{p} = (\theta, K)$.

E.1 NONPERSISTENT POINTS

E.1.1 Simple Bifurcation Points and Isola Centers

Simple bifurcation points and isola centers are special points in the space of $\mathbf{p}$ that satisfy

$$G = \frac{\partial G}{\partial A} = \frac{\partial G}{\partial I} = 0.$$

Because the result will not rely on the assumption $K_1 = K_2 = K$, the symbols K_1 and K_2 are retained during this part of the analysis. Because $\partial G/\partial I = A^2 + (K_2 - 1)A - K_2$, the condition $\partial G/\partial I = 0$ implies

$$A^2 + (K_2 - 1)A - K_2 = 0,$$

which has two solutions: $A = -K_2$ and $A = 1$. The solution $A = -K_2$ is obviously meaningless ([pAKT] cannot be negative). The solution $A = 1$ is also meaningless, because the pAKT level can approach the total AKT level, but can never equal it. That is, $A = [\text{pAKT}]/A_{\max} < 1$. To further demonstrate that $A = 1$ is meaningless, we substitute $A = 1$ into the equation $G = 0$ and obtain

$$K_1 (\phi - \theta + 1) = 0.$$

Because $K_1 \neq 0$, one must have

$$\theta = \phi + 1.$$

By substituting $A = 1$ and $\theta = \phi + 1$ into the equation $\partial G/\partial A = 0$, one has

$$(K_2 + 1)(\phi + \beta\gamma I) - K_1 = 0.$$

Due to the conditions $K_1 \approx 0$ and $K_2 \approx 0$, the above equation reduces to

$$I \approx -\phi/(\beta\gamma).$$

This negative insulin level is absurd. Therefore, there are no simple bifurcation points and isola centers arising from the master equation, i.e., $\mathscr{B} = \phi$.

E.1.2 Double Limit Points

According to lemma 7.1 of [12], polynomial equations of degree 3 or less do not have double limit points. Because Equation (E.1) is a third degree equation, there are no double limit points in the space of $\mathbf{p}$. That is, $\mathscr{D} = \phi$.

E.1.3 Hysteresis Points

Hysteresis points are special points in the space of $\mathbf{p}$ that satisfy the following normal form equation:

$$G = \frac{\partial G}{\partial A} = \frac{\partial^2 G}{\partial A^2} = 0. \tag{E.2}$$

These points constitute a boundary curve $\theta = \Theta_0(K)$ in the phase diagram K versus θ. Because Equation (E.1) has an additional parameter ϕ, there exists a family of curves $\theta = \Theta_0(K)$, each corresponding to a value of ϕ. For a given ϕ, the curve $\theta = \Theta_0(K)$ is obtained numerically by sweeping the K values. For each K, Equation (E.2) is solved to obtain the corresponding θ; a dot (K, θ) is thus drawn on the phase diagram. In this way, the curve $\theta = \Theta_0(K)$ is traced out.

We found Equation (E.2) may not have a meaningful solution for certain K values. For $0 < K \leq K^*$, there is one and only one meaningful solution. For $K > K^*$, the solution is meaningless because the corresponding insulin level I becomes negative. Therefore, K^* must correspond to the critical condition $I = 0$. By solving

$$G = \frac{\partial G}{\partial A} = \frac{\partial^2 G}{\partial A^2} = I = 0, \tag{E.3}$$

the value of K^* and θ^* can be obtained:

$$K^* = 1/(\phi - 1), \tag{E.4}$$

$$\theta^* = (\phi + 1)/\phi. \tag{E.5}$$

Figure E.1 illustrates several phase diagrams, each corresponding to a value of ϕ. In every phase diagram, the curve $\theta = \Theta_0(K)$ starts from (0, 0) and ends at (K^*, θ^*). For instance, the curve of $\phi = 3$ ends at (0.5, 4/3).

E.2 IRREVERSIBILITY POINTS

Irreversibility points are special points in the space of **p** that satisfy

$$G = \frac{\partial G}{\partial A} = I = 0. \tag{E.6}$$

These points constitute a boundary curve $\theta = \Theta_1(K)$ in the phase diagram K versus θ. Because Equation (E.6) has an additional parameter ϕ, there exists a family of curves $\theta = \Theta_1(K)$, each corresponding to a value of ϕ. For a given ϕ, the curve $\theta = \Theta_1(K)$ is obtained numerically by sweeping the K values. For each K, Equation (E.6) is solved to obtain the corresponding θ; a dot (K, θ) is thus drawn on the phase diagram. In this way, the curve $\theta = \Theta_1(K)$ is traced out.

It turns out that the curves $\theta = \Theta_1(K)$ and $\theta = \Theta_0(K)$ have an intersection and the intersection is precisely (K^*, θ^*), where K^* and θ^* have been expressed in Equations (E.4) and (E.5), respectively. This is not strange, because (K^*, θ^*), as a point satisfying Equation (E.3), must also satisfy Equation (E.6).

Figure E.1 illustrates several K versus θ phase diagrams, each corresponding to a value of ϕ. In every phase diagram, the curve $\theta = \Theta_1(K)$ starts from (1, 0) and ends at (K^*, θ^*), the same point at which the other curve ends.

E.3 PHASE DIAGRAM

The intersection of the two boundary curves $\theta = \Theta_0(K)$ and $\theta = \Theta_1(K)$ implies that the region of toggle switches can become a closed area. Figure E.1 shows the phase diagram for different ϕ values. For all the cases, the intersection of the two boundary curves occurs precisely at $K^* = 1/(\phi - 1)$ and $\theta^* = (\phi + 1)/\phi$. For the case $\phi = 3$, one obtains $K^* = 0.5$ and $\theta^* = 1.333$, which is exactly where the two curves intersect in Figure E.1(f). For the case $\phi = 2$, one obtains $K^* = 1$ and $\theta^* = 1.5$, which is exactly where the two curves intersect in Figure E.1(d). For the case $\phi = 1.5$, the two curves intersect at $K^* = 2$ and $\theta^* = 5/3$, but the intersection is out of sight of Figure E.1(c). For the case $\phi = 1.0$, one has $K^* = \infty$, which implies that the two curves do not intersect in Figure E.1(b). For the case $\phi = 0.5$, the two curves do not intersect in Figure E.1(a) because $K^* = -2$ is meaningless.

To see more clearly the effects of ϕ, the boundary curves $\theta = \Theta_0(K)$ and $\theta = \Theta_1(K)$ of $\phi = 2.0, 2.5, 3.0, 3.5$, and 6.0 are superimposed on the same phase diagram

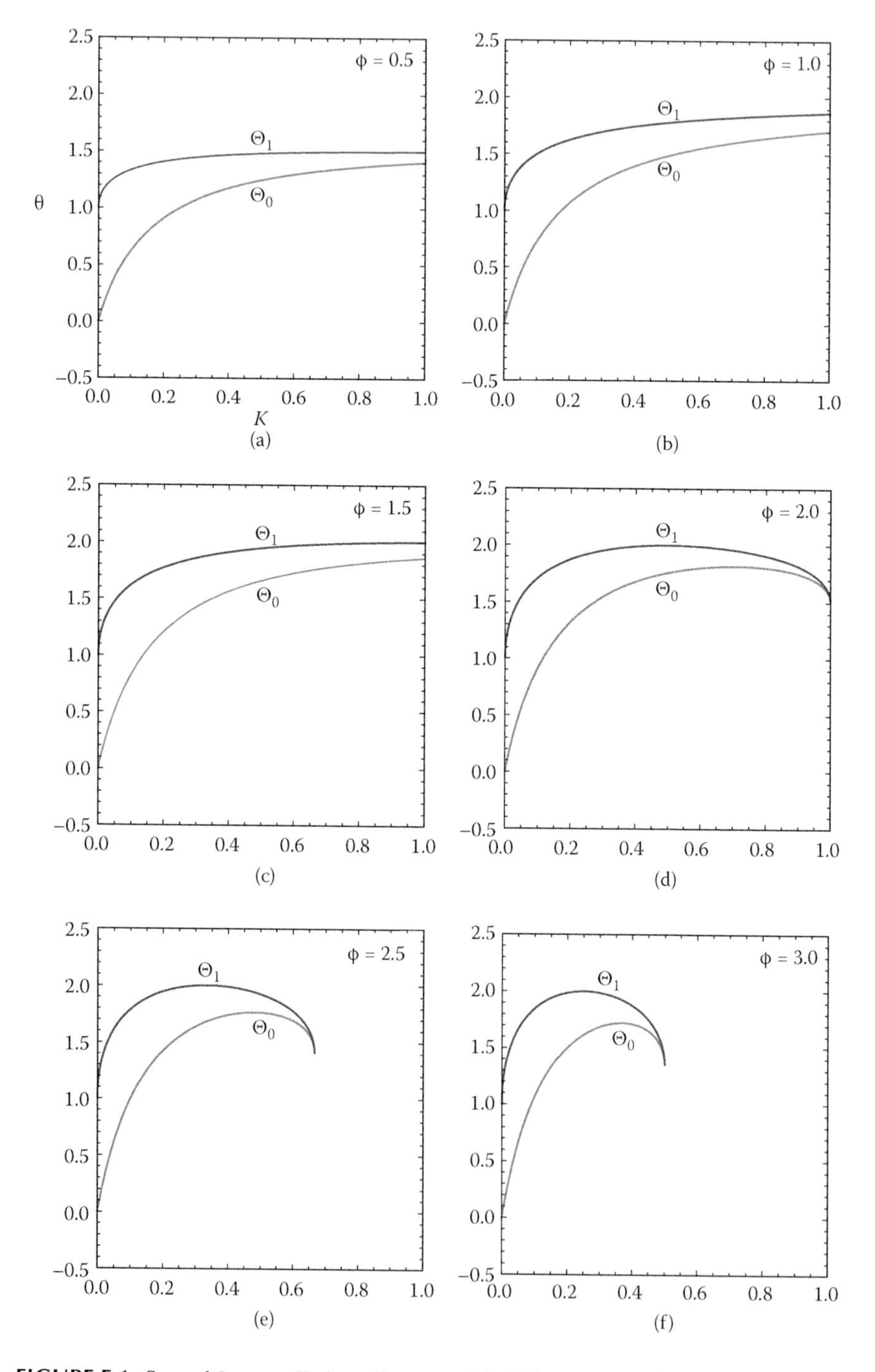

FIGURE E.1 Several θ versus K phase diagrams of the full master equation of AKT activation. Each phase diagram corresponds to a ϕ value.

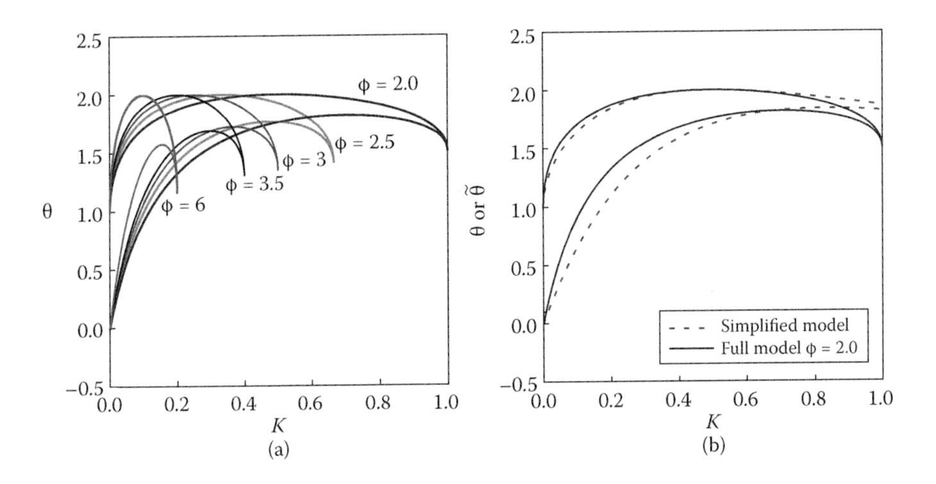

FIGURE E.2 (a) The comparison of θ versus K diagrams of $\phi = 2.0, 2.5, 3.0, 3.5, 6.0$. (b) The comparison between the $\tilde{\theta}$ versus K diagram of the simplified master equation and the θ versus K diagram of the full master equation when $\phi = 2.0$.

(Figure E.2(a)). One sees that the parameter ϕ exerts a great influence on the boundary curves only when K is large. Because the normal K value should be small, the parameter ϕ does not significantly affect the biologically relevant region of the phase diagram. It is the net strength of positive and negative feedback (the parameter θ), rather than the strength of the positive feedback (the parameter ϕ), that primarily determines the boundaries of phase transitions.

Nevertheless, the value of ϕ must be maintained within a suitable range. It is apparent that ϕ cannot be too small (the formation of a toggle switch requires sufficient positive feedback). It was not apparent that ϕ cannot be too large. In the simplified model, the positive feedback can be directly counteracted by the negative feedback, and therefore there is no upper limit for either feedback. In the full model, an overly large ϕ causes an overly small K^* because $K^* = 1/(\phi - 1)$, and thus an overly small region of toggle switches. To obtain a toggle switch, K must be even smaller than this overly small K^*, and thus $A_{\max}$ must be extremely large, which might be biologically unrealistic. For the limit case $\phi \to \infty$, one has $K^* \to 0$ and the region of toggle switches vanishes. The range $2 < \phi < 6$, for example, would be a suitable range for ϕ.

E.4 VALIDITY OF SIMPLIFICATION

Is the simplified master equation a good approximation of the master equation? To answer this question, the boundary curves $\tilde{\Theta}_0(K)$ and $\tilde{\Theta}_1(K)$ of the simplified model (the dashed lines) are compared with the boundary curves $\Theta_0(K)$ and $\Theta_1(K)$ of the full model with $\phi = 2.0$ (the solid lines). One sees that the two phase diagrams differ only slightly. Therefore, the simplified master equation is a good approximation of the full master equation, at least for the case of $\phi = 2.0$. For the other suitable ϕ values, we have demonstrated that they do not perturb significantly the biologically

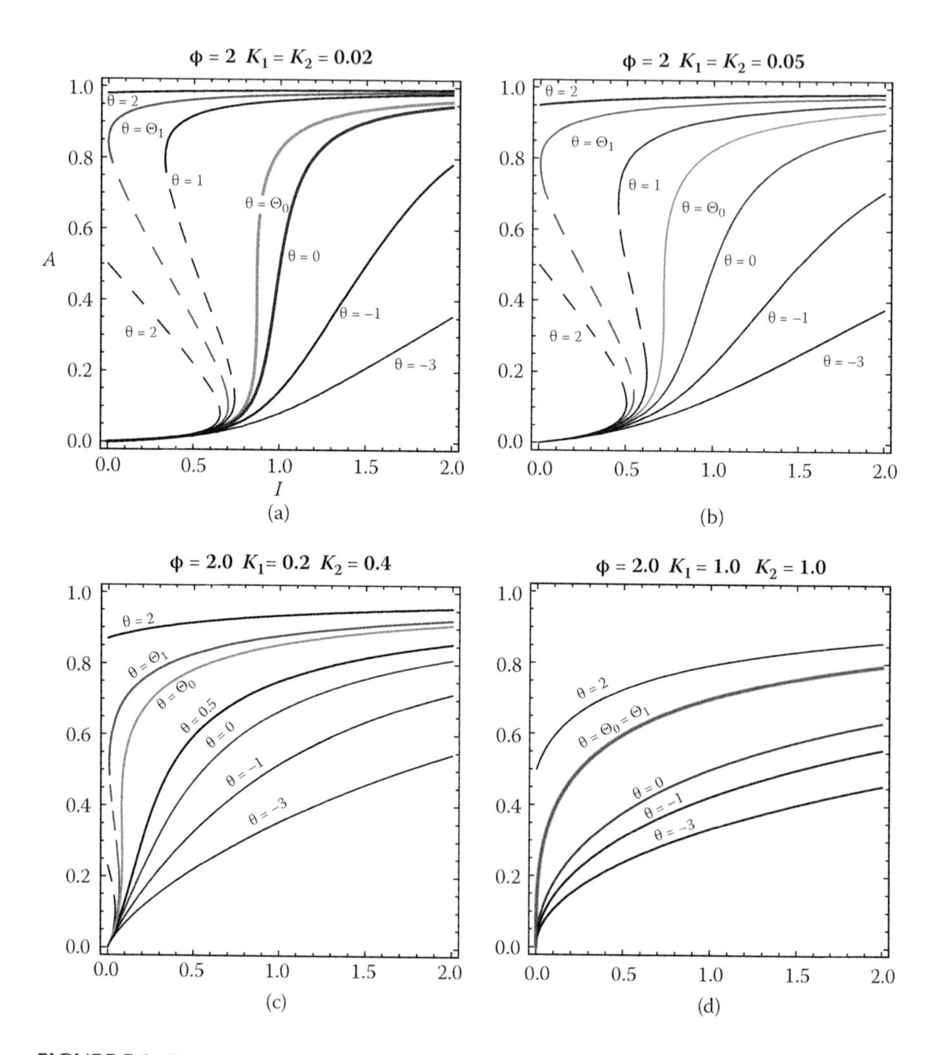

FIGURE E.3 Response curves $A\ (I)$ of the full model with $\phi = 2.0$ and with different K_1, K_2, and θ values.

relevant part of the phase diagram. Therefore, the simplified master equation is a good approximation of the full master equation, for a large range of ϕ values. The analytical results obtained from the simplified model are reliable.

The conclusion that the simplified model approximates well the full model is also reflected by the similarity between the response curves generated by the two models. Figure E.3 illustrates the response curves $A(I)$ of the full model with $\phi = 2.0$, which are similar to those of the simplified model (Figure E.4). Note that in Figure E.3(d) (the case $K_1 = K_2 = 1$), the response curve $A(I)$ corresponding to $\theta = \Theta_0$ is exactly the same as the response curve $A(I)$ corresponding to $\theta = \Theta_1$. This is in line with the fact that in Figure E.2(a), the boundary curve $\theta = \Theta_0(K)$ ($\phi = 2.0$) intersects the boundary curve $\theta = \Theta_1(K)$ ($\phi = 2.0$) at $K^* = 1$.

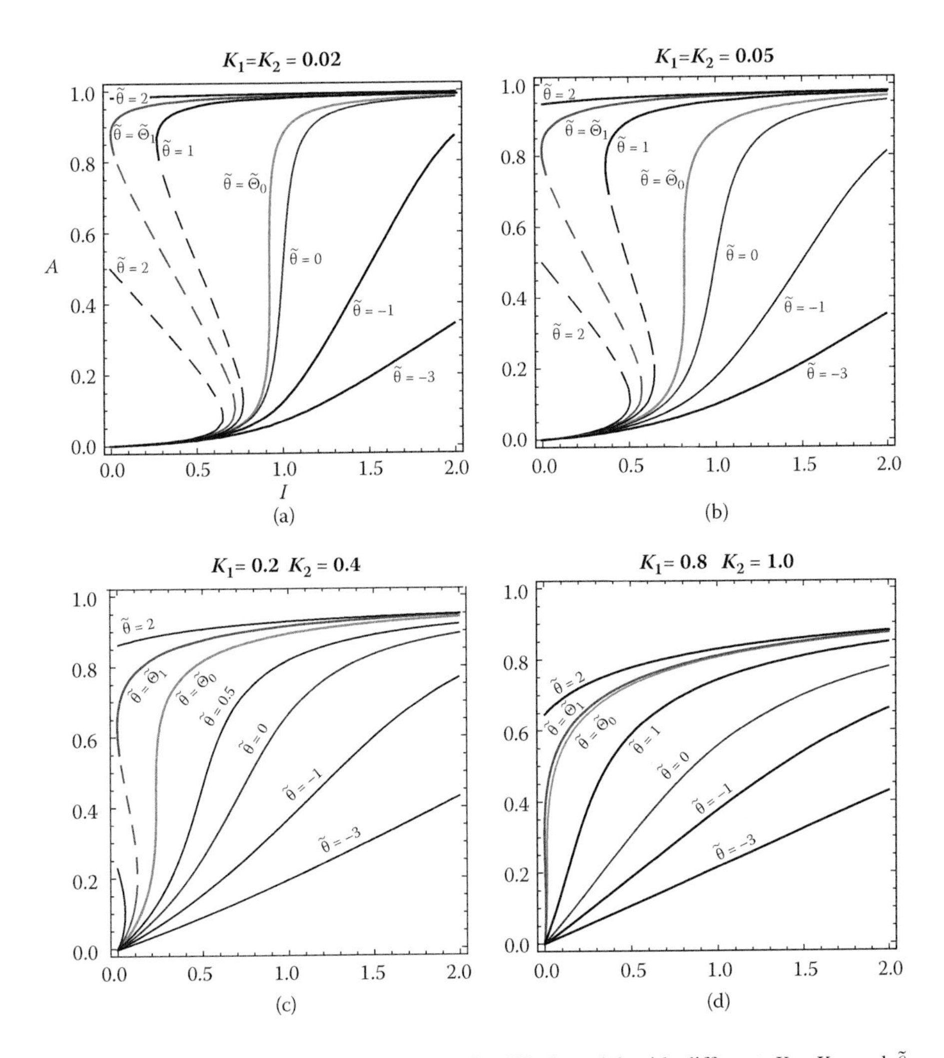

FIGURE E.4 Response curves $A(I)$ of the simplified model with different K_1, K_2, and $\tilde{\theta}$ values.

The similarity between the response curves of the simplified model and those of the full model would pertain to other suitable values of ϕ, if one considers only the biologically relevant range of K.

References

1. Alon, U. 2003. Biological networks: the tinkerer as an engineer. *Science* 301: 1866–1867.
2. Barbour, L.A., J. Shao, L. Qiao, W. Leitner, M. Anderson, J.E. Friedman, and B. Draznin. 2004. Human placental growth hormone increases expression of the p85 regulatory unit of phosphatidylinositol 3-kinase and triggers severe insulin resistance in skeletal muscle. *Endocrinology* 145: 1144–1150.
3. Bergman, R.N., L.S. Phillips, and C. Cobelli. 1981. Physiologic evaluation of factors controlling glucose tolerance in man: measurement of insulin sensitivity and beta-cell glucose sensitivity from the response to intravenous glucose. *J Clin Invest* 68: 1456–1467.
4. Briggs, G.E. and J.B. Haldane. 1925. A note on the kinetics of enzyme action. *Biochem J* 19: 338–339.
5. Cannon, W.B. 1924. *Biographical Memoir, Henry Pickering Bowditch, 1840–1911.* Washington, DC: National Academy of Sciences.
6. Cannon, W.B. 1932. *Wisdom of the Body*. New York: W. W. Norton and Company, Inc.
7. Cinquin, O. and J. Demongeot. 2002. Positive and negative feedback: striking a balance between necessary antagonists. *J Theor Biol* 216: 229–241.
8. Crick, F. 1970. Central dogma of molecular biology. *Nature* 227: 561–563.
9. Ferrell, J.E., Jr. and E.M. Machleder. 1998. The biochemical basis of an all-or-none cell fate switch in *Xenopus oocytes*. *Science* 280: 895–898.
10. Fox, S. 2007. *Human Physiology*, 10th ed. Dubuque, IA: McGraw-Hill Science/Engineering/Math.
11. Goldbeter, A. and D.E. Koshland. 1981. An amplified sensitivity arising from covalent modification in biological systems. *Proc Natl Acad Sci USA* 78: 6840–6844.
12. Golubitsky, M. and D.G. Schaeffer. 1985. *Singularities and Groups in Bifurcation Theory*. New York: Springer.
13. Griffin, M.E., M.J. Marcucci, G.W. Cline, K. Bell, N. Barucci, D. Lee, L.J. Goodyear, E.W. Kraegen, M.F. White, and G.I. Shulman. 1999. Free fatty acid induced insulin resistance is associated with activation of protein kinase C θ and alterations in the insulin signaling cascade. *Diabetes* 48: 1270–1274.
14. Hales, C.N., Barker, D.J.P. 1992. Type 2 (non-insulin-dependent) diabetes mellitus: the thrifty phenotype hypothesis. *Diabetologia* 35:595–601.
15. Hanahan, D. and R.A. Weinberg. 2011. Hallmarks of cancer: the next generation. *Cell* 144: 646–674.
16. Hasty, J., D. McMillen, and J.J. Collins. 2002. Engineered gene circuits. *Nature* 420: 224–230.
17. Isalan, M., C. Lemerle, K. Michalodimitrakis, C. Horn, P. Beltrao, E. Raineri, M. Garriga-Canut, and L. Serrano. 2008. Evolvability and hierarchy in rewired bacterial gene networks. *Nature* 452: 840–845.
18. Kiens, B. 2006. Skeletal muscle lipid metabolism in exercise and insulin resistance. *Physiol Rev* 86: 205–243.
19. Maher, F., S.J. Vannucci, and I.A. Simpson. 1993. Glucose transporter isoforms in brain: absence of GLUT3 from the blood-brain barrier. *J Cereb Blood Flow Metab* 13: 342–345.

20. Malaisse, W., F. Malaisse-Lagae, and P.H. Wright. 1967. A new method for the measurement in vitro of pancreatic insulin secretion. *Endocrinology* 80: 99–108.
21. Mangan, S. and U. Alon. 2003. Structure and function of the feed-forward loop network motif. *Proc Natl Acad Sci USA* 100: 11980–11985.
22. Menten, L. and M.I. Michaelis. 1913. Die Kinetik der Invertinwirkung. *Biochem Z* 49: 333–369.
23. Neel, J.V. 1962. Diabetes mellitus: A "thrifty" genotype rendered detrimental by "progress"? *Am J Hum Genet* 14:355–62.
24. Polonsky, K.S., B.D. Given, and E. Van Cauter. 1988. Twenty-four-hour profiles and pulsatile patterns of insulin secretion in normal and obese subjects. *J Clin Invest* 81: 442–448.
25. Pontryagin, L.S., V.G. Boltyanskii, R.V. Gamkrelidze, and E.F. Mishchenko. 1963. *The Mathematical Theory of Optimal Processes*. New York: Wiley-Interscience.
26. Porte, D., R.S. Sherwin, and A. Baron. 2002. *Ellenberg and Rifkin's Diabetes Mellitus*, 6th ed. New York: McGraw-Hill Professional.
27. Radzicka, A. and R. Wolfenden. 1995. A proficient enzyme. *Science* 267: 90–93.
28. Sha, W., J. Moore, K. Chen, A.D. Lassaletta, C.S. Yi, J.J. Tyson, and J.C. Sible. 2003. Hysteresis drives cell-cycle transitions in *Xenopus laevis* egg extracts. *Proc Natl Acad Sci USA* 100: 975–980.
29. Speakman J. 2008. Thrifty genes for obesity, an attractive but flawed idea, and an alternative perspective: the 'drift gene' hypothesis. *International Journal of Obesity* 32:1611–1617.
30. Sturis, J., K.S. Polonsky, E. Mosekilde, and E. Van Cauter. 1991. Computer model for mechanisms underlying ultradian oscillations of insulin and glucose. *Am J Physiol* 260: E801–E809.
31. Sutherland, W.J. 2005. The best solution. *Nature* 435: 569.
32. Toffolo, G., R.N. Bergman, D.T. Finegood, C.R. Bowden, and C. Cobelli. 1980. Quantitative estimation of beta cell sensitivity to glucose in the intact organism: a minimal model of insulin kinetics in the dog. *Diabetes* 29: 979–990.
33. Topp, B., K. Promislow, G. deVries, R.M. Miura, and D.T. Finegood. 2000. A model of beta-cell mass, insulin, and glucose kinetics: pathways to diabetes. *J Theor Biol* 206: 605–619.
34. Um, S.H., F. Frigerio, M. Watanabe, F. Picard, M. Joaquin, M. Sticker, S. Fumagalli, P.R. Allegrini, S.C. Kozma, J. Auwerx, and G. Thomas. 2004. Absence of S6K1 protects against age- and diet-induced obesity while enhancing insulin sensitivity. *Nature* 431: 200–205.
35. Varlamov, O., R. Somwar, A. Cornea, P. Kievit, K.L. Grove, and C.T. Roberts. 2010. Single-cell analysis of insulin-regulated fatty acid uptake in adipocytes. *Am J Physiol Endocrinol Metab* 299: 486–496.
36. Wang, G., C. Du, H. Chen, R. Simha, Y. Rong, Y. Xiao, and C. Zeng. 2010. Process based network decomposition reveals backbone motif structure. *Proc Natl Acad Sci USA* 107: 10478–10483.
37. Xiong, W. and J.E. Ferrell, Jr. 2003. A positive-feedback-based bistable "memory module" that governs a cell fate decision. *Nature* 426: 460–465.
38. Zhang, J., L.L. Zhang, L. Shen, X.M. Xu, and H.G. Yu. 2013. Regulation of AKT gene expression by cisplatin. *Oncol Lett* 5: 756–760.

Index

K

L

M

N

O

P